MW00490087

Authors
Ted H. Hull, Ed.D.
Ruth Harbin Miles, Ed.S.
Don S. Balka. Ph.D.

Consultant

Don W. Scheuer, Jr., M.S.Ed.
Mathematics Specialist
The Haverford School (ret.)

Publishing Credits

Robin Erickson, *Production Director;*
Lee Aucoin, *Creative Director;* Tim J. Bradley, *Illustration Manager;*
Sara Johnson, M.S.Ed., *Editorial Director;* Maribel Rendón, M.A.Ed., *Editor;*
Jennifer Viñas, *Editor;* Grace Alba, *Designer;*
Corinne Burton, M.A.Ed., *Publisher*

Image Credits

All images Shutterstock

Standards

© 2007 Teachers of English to Speakers of Other Languages, Inc. (TESOL)
© 2007 Board of Regents of the University of Wisconsin System. World-Class Instructional Design and Assessment (WIDA). For more information on using the WIDA ELP Standards, please visit the WIDA website at www.wida.us.
© 2010 National Governors Association Center for Best Practices and Council of Chief State School Officers (CCSS)

Shell Education
5301 Oceanus Drive
Huntington Beach, CA 92649-1030
http://www.shelleducation.com

ISBN 978-1-4258-1291-1
© 2014 Shell Educational Publishing, Inc.

The classroom teacher may reproduce copies of materials in this book for classroom use only. The reproduction of any part for an entire school or school system is strictly prohibited. No part of this publication may be transmitted, stored, or recorded in any form without written permission from the publisher.

Table of Contents

Table of Contents *(cont.)*

Domain: Number and Operations—Fractions

Domain: Measurement and Data

Domain: Geometry

Appendices

Importance of Games

Students learn from play. Play begins when we are infants and continues through adulthood. Games are motivational and educational (Hull, Harbin Miles, and Balka 2013; Burns 2009). They can assist and encourage students to operate as learning communities by requiring students to work together by following rules and being respectful. Games also foster students' thinking and reasoning since students formulate winning strategies. They provide much more sustained practices than do worksheets, and students are more motivated to be accurate. Worksheets may provide 20 to 30 opportunities for students to practice a skill, while games far exceed such prescribed practice opportunities. Lastly, games provide immediate feedback to students concerning their abilities.

Games must be part of the overall instructional approach that teachers use because successful learning requires active student engagement (Hull, Harbin Miles, and Balka 2013; National Research Council 2004), and games provide students with the motivation and interest to become highly engaged. Instructional routines need balance between concept development and skill development. They must also balance teacher-led and teacher-facilitated lessons. Students need time to work independently and collaboratively in order to assimilate information, and games can help support this.

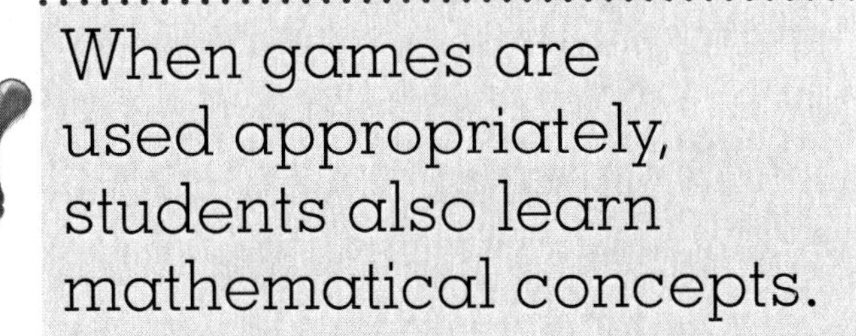

Mathematical Learning

Students must learn mathematics with understanding (NCTM 2000). Understanding means that students know the relationship between mathematical concepts and mathematical skills—mathematical procedures and algorithms work because of the underlying mathematical concepts. In addition, skill proficiency allows students to explore more rigorous mathematical concepts. From this relationship, it is clear that a balance between skill development and conceptual development must exist. There cannot be an emphasis of one over the other.

The National Council of Teachers of Mathematics (2000) and the National Research Council (2001) reinforce this idea. Both organizations state that learning mathematics requires both conceptual understanding and procedural fluency. This means that students need to practice procedures as well as develop their understanding of mathematical concepts in order to achieve success. The games presented in this book reinforce skill-based practice and support students' development of proficiency. These games can also be used as a springboard for discourse about mathematical concepts. The counterpart to this resource is *Math Games: Getting to the Core of Conceptual Understanding*, which builds students' conceptual understanding of mathematics through games.

Importance of Games *(cont.)*

The *Common Core State Standards for Mathematics* (2010) advocate a balanced mathematics curriculum by focusing standards both on mathematical concepts and skills. This is also stressed in the Standards for Mathematical Practice, which discuss the process of "doing" mathematics and the habits of mind students need to possess in order to be successful.

The Standards for Mathematical Practice also focus on the activities that foster thinking and reasoning in which students need to be involved while learning mathematics. Games are an easy way to initiate students in the development of many of the practices. Each game clearly identifies a Common Core domain, a standard, and a skill, and allows students to practice them in a fun and meaningful way.

Games vs. Worksheets

In all likelihood, many mathematics lessons are skill related and are taught and practiced through worksheets. Worksheets heavily dominate elementary mathematics instruction. They are not without value, but they often command too much time in instruction. While students need to practice skills and procedures, the way to practice these skills should be broadened.

Worksheets generally don't promote thinking and reasoning. They become so mechanical that students cease thinking. They are lulled into a feeling that completing is the goal. This sense of "just completing" is not what the Common Core Standards for Mathematical Practice mean when they encourage students to "persevere in solving problems."

Students need to be actively engaged in learning.

Students need to be actively engaged in learning. While worksheets do serve a limited purpose in skill practice, they also contain many potential difficulties. Problems that can occur include the following:

- **Worksheets are often completed in isolation,** meaning that students who are performing a skill incorrectly most likely practice the skill incorrectly for the entire worksheet. The misunderstanding may not be immediately discovered, and in fact, will most likely not be discovered for several days!
- **Worksheets are often boring to students.** Learning a skill correctly is not the students' goal. Their goal becomes to finish the worksheet. As a result, careless errors are often made, and again, these errors may not be immediately discovered or corrected.

Importance of Games *(cont.)*

- **Worksheets are often viewed as a form of subtle punishment.** While perhaps not obvious, the perceived punishment is there. Students who have mastered the skill and can complete the worksheet correctly are frequently "rewarded" for their efforts with another worksheet while they wait for their classmates to finish. At the same time, students who have not mastered the skill and do not finish the worksheet on time are "rewarded" with the requirement to take the worksheet home to complete, or they finish during another portion of the day, often recess or lunch.
- **Worksheets provide little motivation to learn a skill correctly.** There is no immediate correction for mistakes, and often, students do not really care if a mistake is made. When a game is involved, students want and need to get correct answers.

The *Common Core State Standards for Mathematics*, including the Standards for Mathematical Practice, demand this approach change. These are the reasons teachers and teacher leaders must consciously support the idea of using games to support skill development in mathematics.

How to Use This Book

There are many ways to effectively utilize this book. Teachers, mathematics leaders, and parents may use this book to engage students in fun, meaningful, practical mathematics learning. These games can be used as a way to help students maintain skill proficiency or remind them of particular skills prior to a critical concept lesson. These games may also be useful during tutorial sessions, or during class when students have completed their work.

Games at Home

Parents may use these games to work with their child to learn important skills. The games also provide easier ways for parents to interest their child in learning mathematics rather than simply memorizing facts. In many cases, their child is more interested in listening to explanations than correcting their errors.

Parents want to help their children succeed in school, yet they may dread the frequently unpleasant encounters created by completing mathematics worksheets at home. Families can easily use the games in this book by assuming the role of one of the players. At other times, parents provide support and encouragement as their child engages in the game. In either situation, parents are able to work with their children in a way that is fun, educational, and informative.

Games in the Classroom

During game play, teachers are provided excellent opportunities to assess students' abilities and current skill development. Students are normally doing their best and drawing upon their current understanding and ability to play the games, so teachers see an accurate picture of student learning. Some monitoring ideas for teacher assessment include:

- ➔ Move about the room listening and observing
- ➔ Ask student pairs to explain what they are doing
- ➔ Ask the entire class about the game procedures after play
- ➔ Play the game against the class
- ➔ Draw a small group of students together for closer supervision
- ➔ Gather game sheets to analyze students' proficiencies

Ongoing formative assessment and timely intervention are cornerstones of effective classroom instruction. Teachers need to use every available opportunity to make student thinking visible and to respond wisely to what students' visible thinking reveals. Games are an invaluable instructional tool that teachers need to effectively use.

© Shell Education

How to Use This Book *(cont.)*

Students are able to work collaboratively during game play, thus promoting student discourse and deeper learning. The games can also be used to reduce the amount of time students spend completing worksheets.

Each game in this book is based upon a common format. This format is designed to assist teachers in understanding how the game activities are played and which standards and mathematical skills students will be practicing.

Domain
The domain that students will practice is noted at the beginning of each lesson. Each of the five domains addressed in this series has its own icon.

Get Prepared!
Everything a teacher needs to be prepared for game play is noted in the Get Prepared! section. This includes how many copies are needed as well as other tasks that need to be completed with the materials.

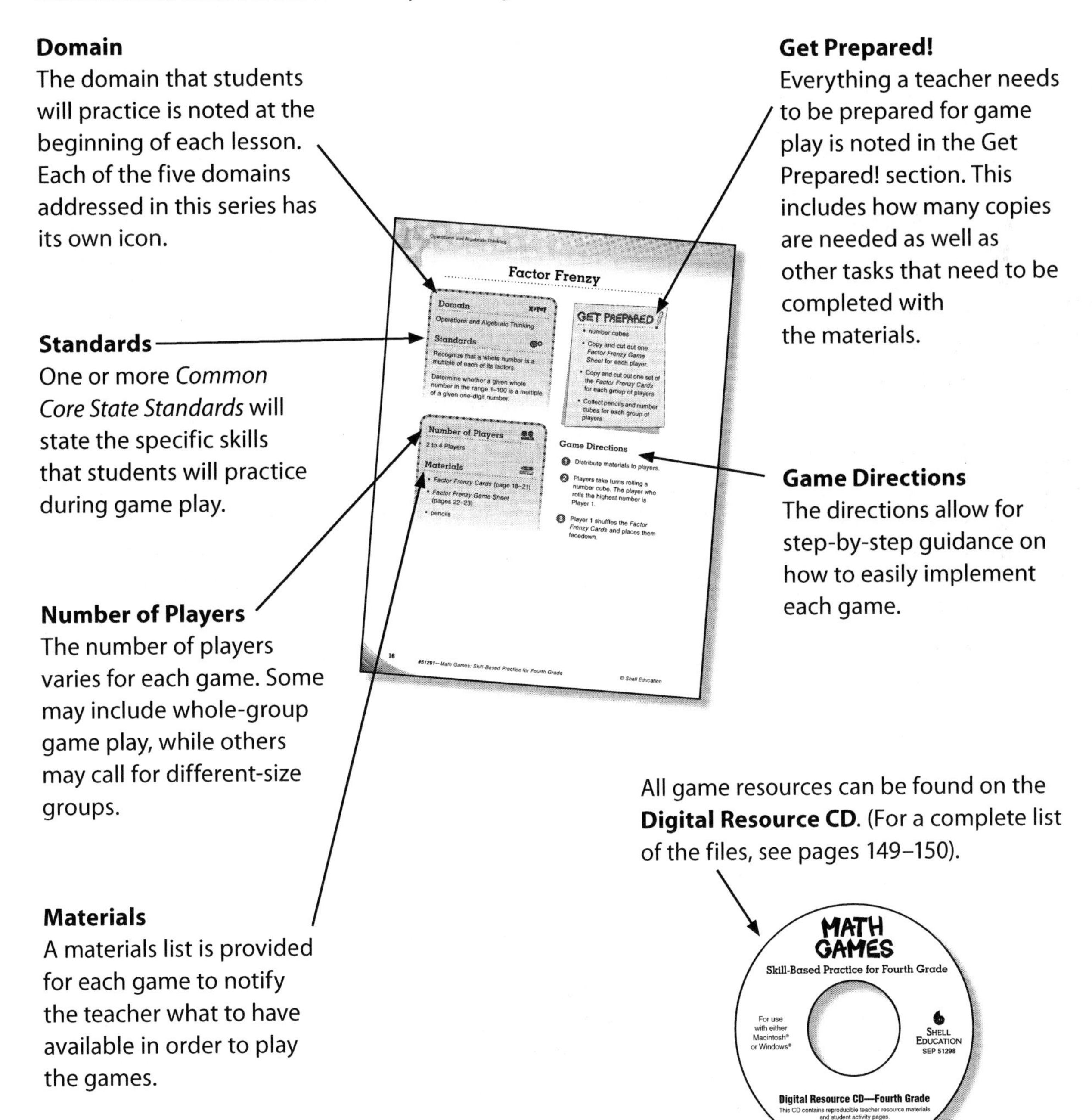

Standards
One or more *Common Core State Standards* will state the specific skills that students will practice during game play.

Game Directions
The directions allow for step-by-step guidance on how to easily implement each game.

Number of Players
The number of players varies for each game. Some may include whole-group game play, while others may call for different-size groups.

Materials
A materials list is provided for each game to notify the teacher what to have available in order to play the games.

All game resources can be found on the **Digital Resource CD**. (For a complete list of the files, see pages 149–150).

How to Use This Book *(cont.)*

Many games include materials such as game boards, activity cards, score cards, and spinners. You may wish to laminate materials for durability.

Game Boards

Some game boards spread across multiple book pages in order to make them larger for game play. When this is the case, cut out each part of the game board and tape them together. Once you cut them apart and tape them together, you may wish to glue them to a large sheet of construction paper and laminate them for durability.

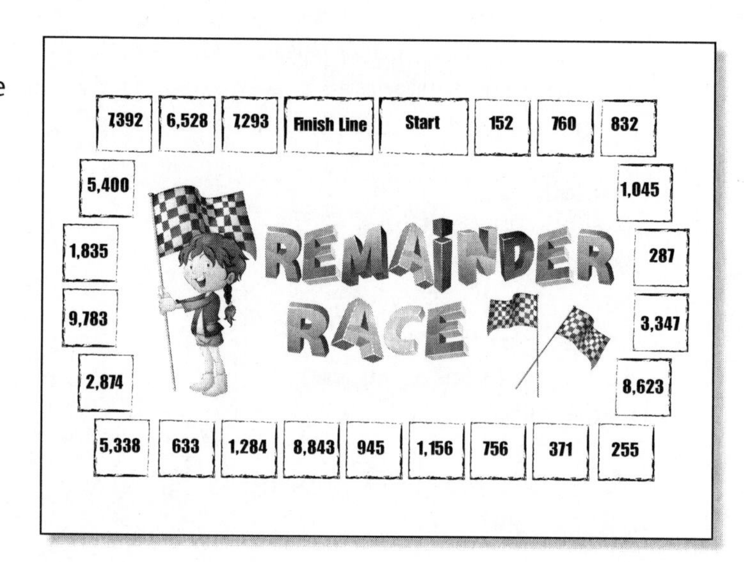

Activity Cards

Some games include activity cards. Once you cut them apart, you may wish to laminate them for durability.

Spinners

Some games include spinners. To use a spinner, cut it out from the page. Place the tip of a pencil in the center with a paperclip around it. Use your other hand to flick the other side of the paperclip.

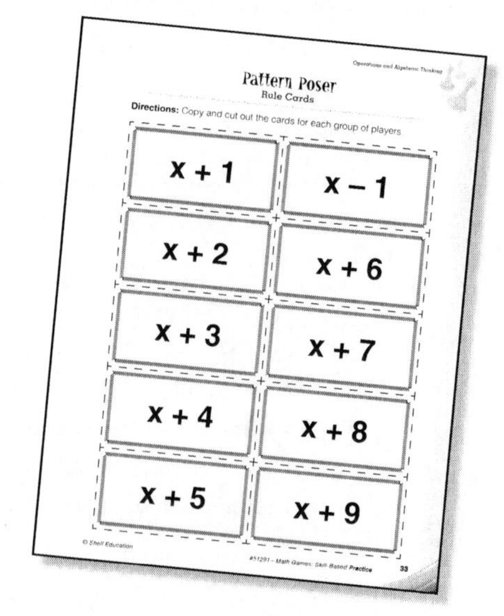
Pattern Poser
Rule Cards

Directions: Copy and cut out the cards for each group of players.

x + 1	x − 1
x + 2	x + 6
x + 3	x + 7
x + 4	x + 8
x + 5	x + 9

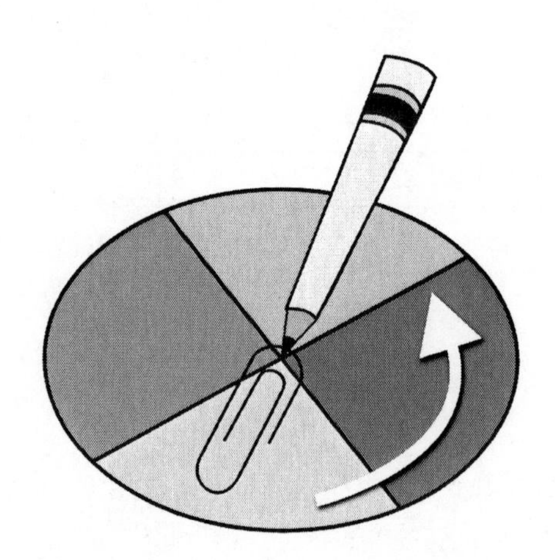

 © Shell Education

Correlation to the Standards

Shell Education is committed to producing educational materials that are research and standards based. In this effort, we have correlated all of our products to the academic standards of all 50 United States, the District of Columbia, the Department of Defense Dependent Schools, and all Canadian provinces.

How to Find Standards Correlations

To print a customized correlation report of this product for your state, visit our website at **http://www.shelleducation.com** and follow the on-screen directions. If you require assistance in printing correlation reports, please contact Customer Service at 1-877-777-3450.

Purpose and Intent of Standards

Legislation mandates that all states adopt academic standards that identify the skills students will learn in kindergarten through grade twelve. Many states also have standards for Pre–K. This same legislation sets requirements to ensure the standards are detailed and comprehensive.

Standards are designed to focus instruction and guide adoption of curricula. Standards are statements that describe the criteria necessary for students to meet specific academic goals. They define the knowledge, skills, and content students should acquire at each level. Standards are also used to develop standardized tests to evaluate students' academic progress. Teachers are required to demonstrate how their lessons meet state standards. State standards are used in the development of all of our products, so educators can be assured they meet the academic requirements of each state.

Common Core State Standards

Many games in this book are aligned to the Common Core State Standards. The standards support the objectives presented throughout the lessons and are provided on the Digital Resource CD (standards.pdf).

TESOL and WIDA Standards

The lessons in this book promote English language development for English language learners. The standards listed on the Digital Resource CD (standards.pdf) support the language objectives presented throughout the lessons.

Standards Correlation Chart

Standard	Game(s)
4.OA.B4—Find all factor pairs for a whole number in the range 1–100. Recognize that a whole number is a multiple of each of its factors. Determine whether a given whole number in the range 1–100 is a multiple of a given one-digit number. Determine whether a given whole number in the range 1–100 is prime or composite.	Factor Frenzy (p. 16); Factor Pairs (p. 24)
4.OAC.5—Generate a number or shape pattern that follows a given rule. Identify apparent features of the pattern that were not explicit in the rule itself. *For example, given the rule "Add 3" and the starting number 1, generate terms in the resulting sequence and observe that the terms appear to alternate between odd and even numbers. Explain informally why the numbers will continue to alternate in this way.*	Pattern Poser (p. 31)
4.NBT.A.2—Read and write multi-digit whole numbers using base-ten numerals, number names, and expanded form. Compare two multi-digit numbers based on meanings of the digits in each place, using >, =, and < symbols to record the results of comparisons.	Standing Order (p. 49); What's My Name? (p. 68)
4.NBT.A.—Use place value understanding to round multi-digit whole numbers to any place.	Spin and Round (p. 43)
4.NBT.B.5—Multiply a whole number of up to four digits by a one-digit whole number, and multiply two two-digit numbers, using strategies based on place value and the properties of operations. Illustrate and explain the calculation by using equations, rectangular arrays, and/or area models.	Multiplication Sprint (p. 35)
4.NBT.B.6—Find whole-number quotients and remainders with up to four-digit dividends and one-digit divisors, using strategies based on place value, the properties of operations, and/or the relationship between multiplication and division. Illustrate and explain the calculation by using equations, rectangular arrays, and/or area models.	Remainder Race (p. 63)

© Shell Education

Standards Correlation Chart *(cont.)*

Standard	Game(s)
4.NF.A.1—Explain why a fraction $\frac{a}{b}$ is equivalent to a fraction $\frac{n \times a}{n \times b}$ by using visual fraction models, with attention to how the number and size of the parts differ even though the two fractions themselves are the same size. Use this principle to recognize and generate equivalent fractions.	Equivalent Fractions Bingo (p. 97)
4.NF.A.2—Compare two fractions with different numerators and different denominators, e.g., by creating common denominators or numerators, or by comparing to a benchmark fraction such as $\frac{1}{2}$. Recognize that comparisons are valid only when the two fractions refer to the same whole. Record the results of comparisons with symbols >, =, or <, and justify the conclusions, e.g., by using a visual fraction model.	Fraction Flip (p. 85); Fraction Action Race (p. 102)
4.NF.B.3b—Decompose a fraction into a sum of fractions with the same denominator in more than one way, recording each decomposition by an equation. Justify decompositions, e.g., by using a visual fraction model. *Examples:* $\frac{3}{8} + \frac{1}{8} + \frac{1}{8} + \frac{1}{8}$; $\frac{3}{8} + \frac{1}{8} + \frac{1}{8}$; $2\frac{1}{8} = 1 + 1 + \frac{1}{8} = \frac{8}{8} + \frac{8}{8} + \frac{1}{8}$.	Fishing for Fractions (p. 76)
4.NF.B.3c—Add and subtract mixed numbers with like denominators, e.g., by replacing each mixed number with an equivalent fraction, and/or by using properties of operations and the relationship between addition and subtraction.	Improper Fraction Bingo (p. 107)
4.NF.B.4a—Understand a fraction $\frac{a}{b}$ as a multiple of $\frac{1}{b}$. *For example, use a visual fraction model to represent* $\frac{5}{4}$ *as the product* $5 \times (\frac{1}{4})$, *recording the conclusion by the equation* $\frac{5}{4} = 5 \times (\frac{1}{4})$.	Fishing for Fractions (p. 76)
4.NF.C.6—Use decimal notation for fractions with denominators 10 or 100. *For example, rewrite 0.62 as* $\frac{62}{100}$; *describe a length as 0.62 meters; locate 0.62 on a number line diagram.*	Fraction Decimal Match (p. 91)
4.NF.C.7—Compare two decimals to hundredths by reasoning about their size. Recognize that comparisons are valid only when the two decimals refer to the same whole. Record the results of comparisons with the symbols >, =, or <, and justify the conclusions, e.g., by using a visual model.	Dueling Decimals (p. 81)

Standards Correlation Chart *(cont.)*

Standard	Game(s)
4.MD.A.1—Know relative sizes of measurement units within one system of units, including km, m, cm; kg, g; lb, oz; l, ml; hr, min, sec. Within a single system of measurement, express measurements in a larger unit in terms of a smaller unit. Record measurement equivalents in a two-column table. *For example, know that 1 ft is 12 times as long as 1 in. Express the length of a 4 ft snake as 48 in. Generate a conversion table for feet and inches listing the number pairs (1, 12), (2, 24), (3, 36), ...*	Pathway Puzzles (p. 112)
4.MD.A.3—Apply the area and perimeter formulas for rectangles in real world and mathematical problems. *For example, find the width of a rectangular room given the area of the flooring and the length, by viewing the area formula as a multiplication equation with an unknown factor.*	Measurement Match (p. 121)
A.G.A.1—Draw points, lines, line segments, rays, angles (right, acute, obtuse), and perpendicular and parallel lines. Identify these in two-dimensional figures.	Geo Identity (p. 127)
4.G.A.2—Classify two-dimensional figures based on the presence or absence of parallel or perpendicular lines, or the presence or absence of angles of a specified size. Recognize right triangles as a category, and identify right triangles.	Classify Bingo (p. 140)
4.G.A.3—Recognize a line of symmetry for a two-dimensional figure as a line across the figure such that the figure can be folded along the line into matching parts. Identify line-symmetric figures and draw lines of symmetry.	Symmetric Shapes (p. 133)

© Shell Education

About the Authors

Ted H. Hull, Ed.D., served in public education for 32 years as a mathematics teacher, a K–12 mathematics coordinator, a school principal, director of curriculum and instruction, and project director for the Charles A. Dana Center at the University of Texas in Austin. While at the University of Texas, he directed the research project "Transforming Schools: Moving from Low-Achieving to High Performing Learning Communities." After retiring, Ted opened LCM: Leadership • Coaching • Mathematics with his coauthors and colleagues. Ted has coauthored numerous books addressing mathematics improvement and has served as the Regional Director for the National Council of Supervisors of Mathematics (NCSM).

Ruth Harbin Miles, Ed.S., currently coaches inner-city, rural, and suburban mathematics teachers and serves on the Board of Directors for the National Council of Teachers of Mathematics, the National Council of Supervisors of Mathematics and Virginia's Council of Mathematics Teachers. Her professional experiences include coordinating the K–12 Mathematics Department for Olathe, Kansas Schools and adjunct teaching for Mary Baldwin College and James Madison University in Virginia. A coauthor of four books on transforming teacher practice through team leadership, mathematics coaching, and visible student thinking and co-owner of Happy Mountain Learning, Ruth's specialty and passion include developing teachers' content knowledge and strategies for engaging students to achieve high standards in mathematics.

Don S. Balka, Ph.D., a former middle school and high school mathematics teacher, is Professor Emeritus in the Mathematics Department at Saint Mary's College in Notre Dame, Indiana. Don has presented at over 2,000 workshops, conferences, and in-service trainings throughout the United States and has authored or coauthored over 30 books on mathematics improvement. Don has served as director for the National Council of Teachers of Mathematics, the National Council of Supervisors of Mathematics, TODOS: Mathematics for All, and the School Science and Mathematics Association. He is currently president of TODOS and past president of the School Science and Mathematics Association.

Factor Frenzy

Domain

X+Y=?

Operations and Algebraic Thinking

Standards

Recognize that a whole number is a multiple of each of its factors.

Determine whether a given whole number in the range 1–100 is a multiple of a given one-digit number.

Number of Players

2 to 4 Players

Materials

- *Factor Frenzy Cards* (pages 18–21)
- *Factor Frenzy Game Sheet* (page 22)
- *Factor Frenzy Grid* (page 23)
- number cubes

GET PREPARED!

- Copy and cut out the *Factor Frenzy Game Sheet* and one *Factor Frenzy Grid* for each player.
- Copy and cut out one set of the *Factor Frenzy Cards* for each group of players.
- Collect a number cube for each group of players.

Game Directions

1. Distribute materials to players.
2. Players take turns rolling a number cube. The player who rolls the highest number is Player 1.
3. Player 1 shuffles the *Factor Frenzy Cards* and places them facedown.

© Shell Education

Factor Frenzy *(cont.)*

4. Player 1 draws a card from the deck. He or she writes the number on the *Factor Frenzy Game Sheet*. The player then lists all the factors of that number on the game sheet. For example, if a card with the number 20 is drawn, the player lists 1, 2, 4, 5, 10, and 20.

5. Player 1 chooses one of the factors and crosses it off on the *Factor Frenzy Grid*. The player also writes the matching multiplication sentence in that same box on the grid. For example: For the number 20, the player might cross off 4 and write $4 \times 5 = 20$ in the box.

6. Player 2 repeats steps 4 and 5.

7. Players take turns drawing cards from the deck, listing factors, and crossing factors off the grid. Players can only cross off each factor once.

8. If a player cannot cross off a factor, he or she loses that turn.

9. The first player to cross off all the factors on his or her *Factor Frenzy Grid* wins.

© Shell Education

Factor Frenzy

Cards

Directions: Copy and cut out one set of cards for each group of players.

© Shell Education

Factor Frenzy

Cards (cont.)

Factor Frenzy

Cards *(cont.)*

© Shell Education

Factor Frenzy

Cards *(cont.)*

© Shell Education

Name:______________________________ Date:________________

Factor Frenzy

Game Sheet

Directions: Record the number drawn and all of its factors on the sheet.

Number	Factors

 © Shell Education

Name: ______________________________ Date: ________________

Factor Frenzy

Grid

Directions: Cross off one factor from the grid and use it in a number sentence.

Factor Grid		
Factor: 1 **Number Sentence:**	**Factor: 2** **Number Sentence:**	**Factor: 3** **Number Sentence:**
Factor: 4 **Number Sentence:**	**Factor: 5** **Number Sentence:**	**Factor: 6** **Number Sentence:**
Factor: 7 **Number Sentence:**	**Factor: 8** **Number Sentence:**	**Factor: 9** **Number Sentence:**

Factor Pairs

Domain

X+Y=?

Operations and Algebraic Thinking

Standard

Find all factor pairs for a whole number in the range 1–100.

Number of Players

2 Players

Materials

- *Factor Pairs Number Cards* (pages 26–28)
- *Factor Pairs Recording Sheet Part 1* (page 29)
- *Factor Pairs Recording Sheet Part 2* (page 30)
- paper bags

GET PREPARED!

- Copy and cut out the *Factor Pairs Recording Sheet Part 1 and Factor Pairs Recording Sheet Part 2* and the *Factor Pairs Number Cards* for each pair of players.
- Choose 10 *Factor Pairs Number Cards* and place them in a paper bag for each pair of players. There are more cards than are needed for play, so not all of them will be used.

Game Directions

1. Distribute materials to players.
2. Play begins when the teacher draws a number card from the paper bag and calls out the number shown.
3. Players work quickly with their partners to write the number on the *Factor Pairs Recording Sheet Part 1* or *Part 2* and list as many factor pairs for the number as possible.

© Shell Education

Factor Pairs *(cont.)*

4. The teacher verifies the correct factor pairs. Pairs of players earn 1 point for each correct factor pair.

   For example:

Number	Factor Pairs	Points
72	1 × 72 2 × 36 3 × 24 4 × 18 6 × 12 8 × 9	6

5. Play continues for 10 rounds. The group with the most points after 10 rounds wins.

Factor Pairs

Number Cards

Directions: Copy and cut out the cards for each pair of players.

 © Shell Education

Factor Pairs

Number Cards *(cont.)*

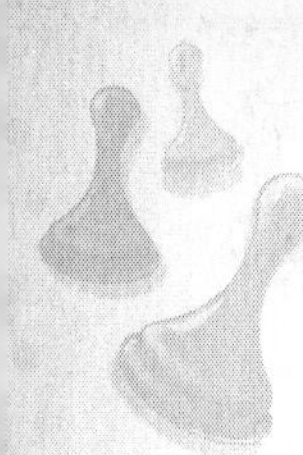

Factor Pairs

Number Cards *(cont.)*

© Shell Education

Name: ______________________ Date: ______________

Factor Pairs

Recording Sheet Part 1

Directions: Record the number on the sheet. Then, list as many factor pairs as you can. One point is earned for each correct factor pair.

Number	Factor Pairs	Points
Sample Round **Example:** *65*	*1* × *65* *5* × *13*	*2*
Round 1		
Round 2		
Round 3		
Round 4		

© Shell Education

Factor Pairs

Recording Sheet Part 2

Number	Factor Pairs	Points
Round 5		
Round 6		
Round 7		
Round 8		
Round 9		
Round 10		
	Total Points	

© Shell Education

Pattern Poser

Domain

Operations and Algebraic Thinking

Standard

Generate a number or shape pattern that follows a given rule.

Number of Players

2 Players

Materials

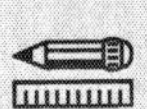

- *Pattern Poser Rule Cards* (page 33)
- *Pattern Poser Recording Sheet* (page 34)
- number cubes

GET PREPARED

- Copy and cut out one *Pattern Poser Recording Sheet* for each player.
- Copy and cut out one set of the *Pattern Poser Rule Cards* for each pair of players.
- Collect a number cube for each pair of players.

Game Directions

1. Distribute materials to players.
2. Players take turns rolling a number cube. The player who rolls the higher number is Player 1.
3. The *Pattern Poser Rule Cards* are shuffled and placed facedown. Player 1 turns over the top card in the deck and lays it face up.

© Shell Education

Pattern Poser *(cont.)*

4. Both players copy the rule from the card onto their recording sheets. For example:

Rule: $x - 1$

5. Player 1 rolls the number cube. He or she substitutes the number rolled for the *x* in the rule and solves the number sentence. For example:

3	$3 - 1 = 2$

6. Player 2 rolls the number cube. He or she substitutes the number rolled for the *x* in the rule and solves the number sentence.

7. Players take turns rolling the number cube and substituting the value rolled for *x* in the number sentence. If a duplicate value is rolled, the player loses his or her turn. After two unsuccessful rolls, a player may select a value for *x*.

8. The first player to fill in every line of the recording sheet—by substituting 1, 2, 3, 4, 5, and 6 for *x*—wins the round.

© Shell Education

Pattern Poser

Rule Cards

Directions: Copy and cut out the cards for each group of players.

$x + 1$	$x - 1$
$x + 2$	$x + 6$
$x + 3$	$x + 7$
$x + 4$	$x + 8$
$x + 5$	$x + 9$

© Shell Education

Name: ____________________ Date: ____________

Pattern Poser

Recording Sheet

Directions: Write number sentences and solve them using the value of *x*.

EXAMPLE

Rule: $x - 1$

Value of *x*	Number Sentence
1	$1 - 1 = 0$
2	$2 - 1 = 1$
3	$3 - 1 = 2$
4	$4 - 1 = 3$
5	$5 - 1 = 4$
6	$6 - 1 = 5$

Round 1

Rule:

Value of *x*	Number Sentence
1	
2	
3	
4	
5	
6	

Round 2

Rule:

Value of *x*	Number Sentence
1	
2	
3	
4	
5	
6	

Round 3

Rule:

Value of *x*	Number Sentence
1	
2	
3	
4	
5	
6	

© Shell Education

Multiplication Sprint

Domain

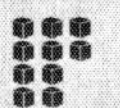

Number and Operations in Base Ten

Standard

Multiply a whole number of up to four digits by a one-digit whole number, and multiply two two-digit numbers, using strategies based on place value and the properties of operations.

Number of Players

2 Players

Materials

- *Multiplication Sprint Game Board* (pages 37–38)
- *Multiplication Sprint Game Markers* (page 39)
- *Multiplication Sprint Recording Sheet* (pages 40–41)
- *Multiplication Sprint Digit Deck* (page 42)
- paper bags
- crayons and scissors

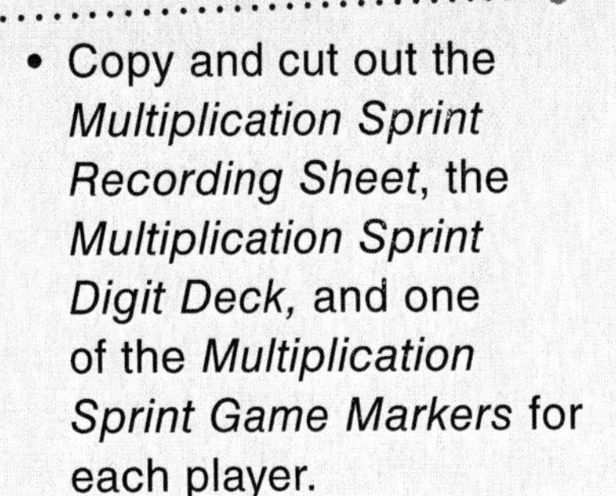

GET PREPARED!

- Copy and cut out the *Multiplication Sprint Recording Sheet*, the *Multiplication Sprint Digit Deck,* and one of the *Multiplication Sprint Game Markers* for each player.
- Copy and cut out the *Multiplication Sprint Game Board* for each pair of players.
- Place the *Multiplication Sprint Digit Deck* in a paper bag for each pair of players.
- Collect crayons and scissors for each pair of players.

Multiplication Sprint *(cont.)*

Game Directions

1. Distribute materials to players. Each player colors the *Multiplication Sprint Game Markers* a different color to distinguish between the two.

2. Each player draws three cards from the bag. Players record the digits on their *Multiplication Sprint Recording Sheet* to make a large two-digit number. The third number becomes the one-digit multiplier. Players multiply to find the products.

3. The player with the greater product moves his or her game marker one space on the *Multiplication Sprint Game Board.*

4. Players return the cards to the bag, shake the bag, and draw three new cards. Play continues until one player reaches the finish line. In the event of a tie, players play one additional round.

 © Shell Education

Multiplication Sprint

Game Board

Directions: Cut out the game board. Tape it to the game board on page 38.

Multiplication

© Shell Education

Multiplication Sprint

Game Board *(cont.)*

 © Shell Education

Multiplication Sprint

Game Markers

Directions: Copy and cut out the game markers. Make sure players use different colors on each marker so the difference between the two is clear. Players will use these to mark their movements on the game board.

© Shell Education

Name:______________________________________ Date:____________________

Multiplication Sprint

Recording Sheet

Directions: Record your digits and multiply them together.

☐☐ × ☐ ______	☐☐ × ☐ ______
☐☐ × ☐ ______	☐☐ × ☐ ______
☐☐ × ☐ ______	☐☐ × ☐ ______
☐☐ × ☐ ______	☐☐ × ☐ ______

© Shell Education

Multiplication Sprint

Recording Sheet *(cont.)*

☐☐ × ☐	☐☐ × ☐
☐☐ × ☐	☐☐ × ☐
☐☐ × ☐	☐☐ × ☐
☐☐ × ☐	☐☐ × ☐

Multiplication Sprint

Digit Deck

Directions: Make one copy for each group. Cut the cards apart and place each 20-card deck in a paper bag.

0	1	2	3
4	5	6	7
8	9	0	1
2	3	4	5
6	7	8	9

© Shell Education

Spin and Round

Domain

Number and Operations in Base Ten

Standard

Use place value understanding to round multi-digit whole numbers to any place.

Number of Players

Groups of 4

Materials

- *Spin and Round Recording Sheet* (page 45)
- *Spin and Round Digit Deck* (pages 46–47)
- *Spin and Round Spinner* (page 48)
- paper bags
- paperclips and sharpened pencils
- number cubes

GET PREPARED!

- Copy and cut out the *Spin and Round Digit Deck* and the *Spin and Round Spinner* for each group of players.
- Copy the *Spin and Round Recording Sheet* for each player.
- Place the *Spin and Round Digit Deck* in a paper bag for each group of players.
- Collect one paperclip and one sharpened pencil for each group to use with spinners.
- Collect one number cube for each group of players.

Game Directions

1. Distribute materials to players.
2. Players take turns rolling a number cube. The player who rolls the lowest number is Player 1.
3. Player 1 begins the game by saying either "greatest number wins" or "least number wins."

Spin and Round *(cont.)*

4. Each player draws six cards from the bag. Players place the cards on their *Spin and Round Recording Sheets* in the six corresponding squares to make the greatest (or least) six-digit number possible, depending on which one Player 1 chooses to win.

5. Player 1 flicks the paperclip to determine to which place all players will round their numbers. Players round their number to the indicated place and record the number on their recording sheet.

6. The player with the greatest (or least) number wins the round. He or she makes a tally mark to record the win on the recording sheet. Players return their cards to the bag.

7. Player 2 begins the next round by saying either "greatest number wins" or "least number wins." Each player draws six cards again, makes a number, and repeats steps 5 and 6.

8. Play continues for eight rounds. The player with the most wins at the end of the game is the winner.

 © Shell Education

Name: ______________________________ Date: ________________

Spin and Round

Recording Sheet

Directions: Round the six-digit numbers and record them on your sheet.

Name: ______________ Date: ______________

My Wins

1. Round to nearest ________: ____________

2. Round to nearest ________: ____________

3. Round to nearest ________: ____________

4. Round to nearest ________: ____________

5. Round to nearest ________: ____________

6. Round to nearest ________: ____________

7. Round to nearest ________: ____________

8. Round to nearest ________: ____________

© Shell Education

Directions: Make a copy and cut out one set of cards for each group of players.

0	1	2	3
4	5	6	7
8	9	0	1
2	3	4	5
6	7	8	9

© Shell Education

Spin and Round

Digit Deck *(cont.)*

0	1	2	3
4	5	6	7
8	9	0	1
2	3	4	5
6	7	8	9

Spinner

Directions: Copy and cut out the spinner for each group of players. For steps on how to assemble this spinner, see page 10.

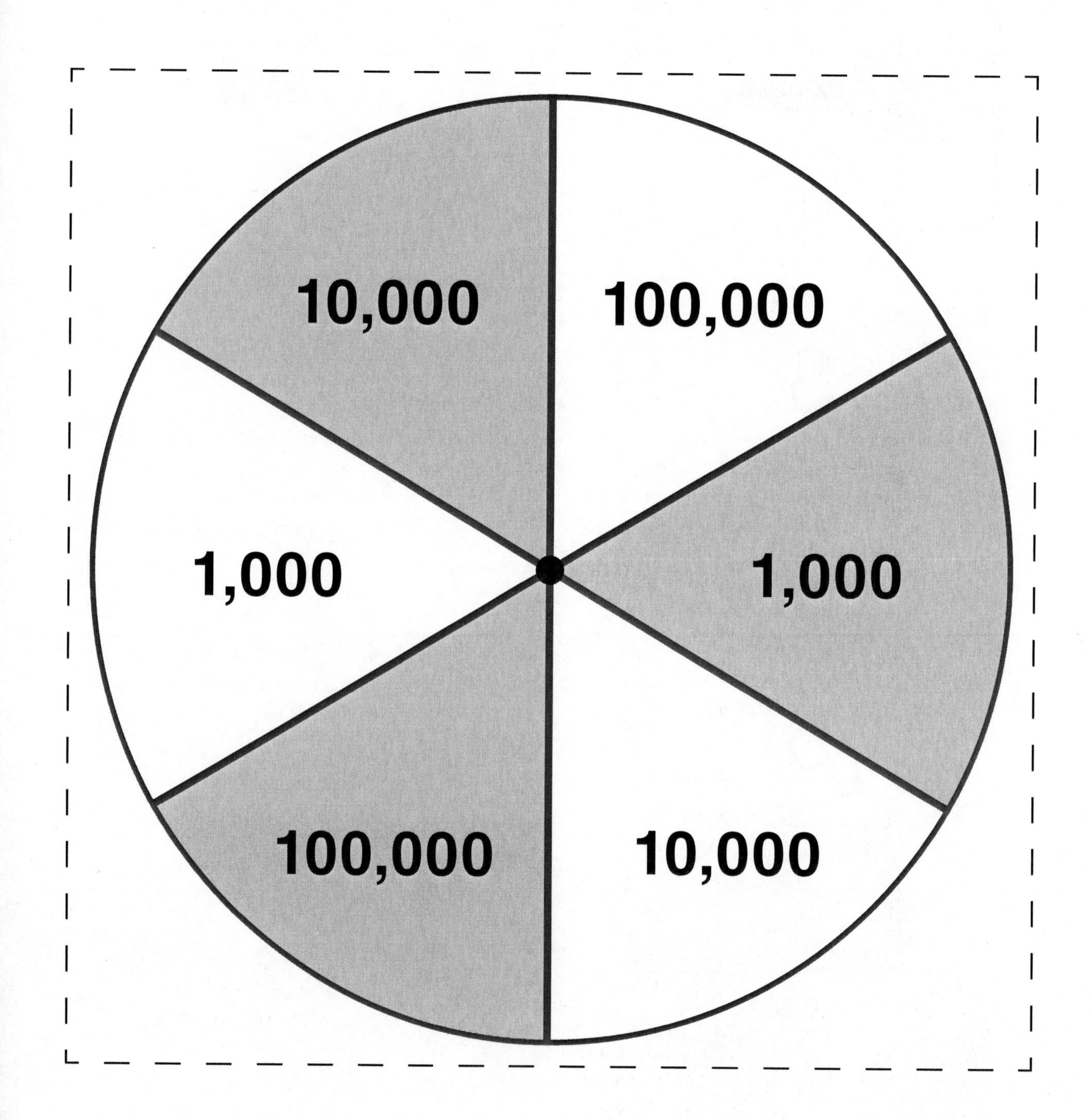

 © Shell Education

Standing Order

Domain

Number and Operations in Base Ten

Standard

Read and write multi-digit whole numbers using base-ten numerals, number names, and expanded form.

Number of Players

Whole class divided into two teams of players

Materials

Standing Order Number Cards, Sets 1–6 (pages 51–62)

GET PREPARED

Copy and cut out the *Standing Order Number Cards, Sets 1–6* for each team.

Game Directions

1. Distribute materials to players.
2. Team 1 receives *Standing Order Number Cards, Set 1,* and Team 2 receives *Standing Order Number Cards, Set 2.*
3. Half of the players on each team receive a card with a multi-digit whole number, and the other half of the team receives cards with an expanded form number.

Standing Order *(cont.)*

4. The game begins when the teacher says, "Go!" Players on each team work to find a match for their number. For example, a player with a card containing 315,741 will find a player with the card that has the matching expanded form number like the sample below.

315,741	300,000 10,000 5,000 700 40 + 1

5. The first team to make all their matches wins the round of play.

6. Repeat play using *Standing Order Number Cards, Sets 3–6*. The teacher keeps track of the rounds on the board. The team with the most rounds wins the game.

© Shell Education

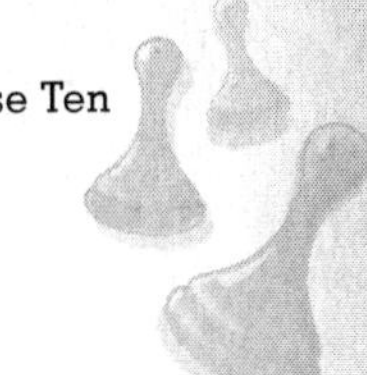

Standing Order

Number Cards Set 1

Directions: Copy and cut out the cards for each team of players.

456,831	**456,813**
456,183	**451,683**
400,000 50,000 6,000 800 30 + 1	400,000 50,000 6,000 800 10 + 3
400,000 50,000 6,000 100 80 + 3	400,000 50,000 1,000 600 80 + 3

© Shell Education

Standing Order

Number Cards Set 1 *(cont.)*

415,683	145,683
486,531	486,153
400,000 10,000 5,000 600 80 + 3	100,000 40,000 5,000 600 80 + 3
400,000 80,000 6,000 500 30 + 1	400,000 80,000 6,000 100 50 + 3

© Shell Education

Standing Order

Number Cards Set 2

Directions: Copy and cut out the cards for each team of players.

485,631	**485,136**
483,651	**483,516**
400,000 80,000 5,000 600 30 + 1	400,000 80,000 5,000 100 30 + 6
400,000 80,000 3,000 600 50 + 1	400,000 80,000 3,000 500 10 + 6

Standing Order

Number Cards Set 2 *(cont.)*

436,851	436,185
385,641	385,416
400,000 30,000 6,000 800 50 + 1	400,000 30,000 6,000 100 80 + 5
300,000 80,000 5,000 600 40 + 1	300,000 80,000 5,000 400 10 + 6

 © Shell Education

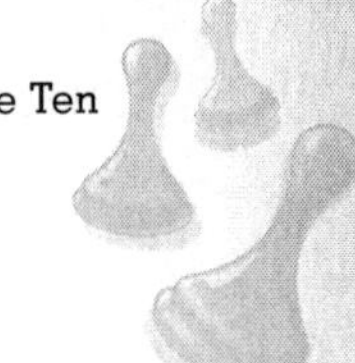

Standing Order

Number Cards Set 3

Directions: Copy and cut out the cards for each team of players.

932,674	**932,647**
932,467	**934,267**
900,000 30,000 2,000 600 70 + 4	900,000 30,000 2,000 600 40 + 7
900,000 30,000 2,000 400 60 + 7	900,000 30,000 4,000 200 60 + 7

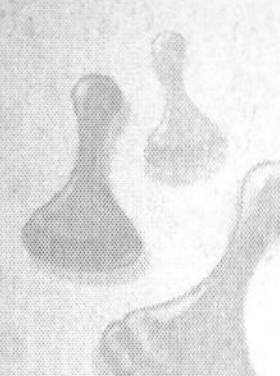

Standing Order

Number Cards Set 3 *(cont.)*

943,267	947,632
947,236	976,234
900,000 40,000 3,000 200 60 + 7	900,000 40,000 7,000 600 30 + 2
900,000 40,000 7,000 200 30 + 6	900,000 70,000 6,000 200 30 + 4

© Shell Education

Standing Order

Number Cards Set 4

Directions: Copy and cut out the cards for each team of players.

976,432	743,629
743,962	639,724
900,000 70,000 6,000 400 30 + 2	700,000 40,000 3,000 600 20 + 9
700,000 40,000 3,000 900 60 + 2	600,000 30,000 9,000 700 20 + 4

Standing Order

Number Cards Set 4 *(cont.)*

639,427	**479,623**
479,362	**326,794**
600,000 30,000 9,000 400 20 + 7	400,000 70,000 9,000 600 20 + 3
400,000 70,000 9,000 300 60 + 2	300,000 20,000 6,000 700 90 + 4

© Shell Education

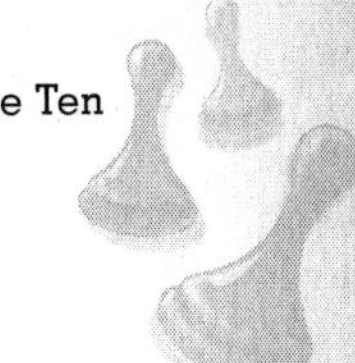

Standing Order

Number Cards Set 5

Directions: Copy and cut out the cards for each team of players.

572,836	572,863
572,683	576,283
500,000 70,000 2,000 800 30 + 6	500,000 70,000 2,000 800 60 + 3
500,000 70,000 2,000 600 80 + 3	500,000 70,000 6,000 200 80 + 3

© Shell Education

Standing Order

Number Cards Set 5 *(cont.)*

567,283

657,283

657,832

687,523

500,000
60,000
7,000
200
80
\+ 3

600,000
50,000
7,000
200
80
\+ 3

600,000
50,000
7,000
800
30
\+ 2

600,000
80,000
7,000
500
20
\+ 3

© Shell Education

Standing Order

Number Cards Set 6

Directions: Copy and cut out the cards for each team of players.

687,235	**756,382**
756,238	**823,567**
600,000 **80,000** **7,000** **200** **30** **+ 5**	**700,000** **50,000** **6,000** **300** **80** **+ 2**
700,000 **50,000** **6,000** **200** **30** **+ 8**	**800,000** **20,000** **3,000** **500** **60** **+ 7**

© Shell Education

Standing Order

Number Cards Set 6 *(cont.)*

823,765	**238,576**
238,675	**385,762**
800,000 20,000 3,000 700 60 + 5	200,000 30,000 8,000 500 70 + 6
200,000 30,000 8,000 600 70 + 5	300,000 80,000 5,000 700 60 + 2

© Shell Education

Remainder Race

Domain

Number and Operations in Base Ten

Standard

Find whole-number quotients and remainders with up to four-digit dividends and one-digit divisors, using strategies based on place value, the properties of operations, and/or the relationship between multiplication and division.

GET PREPARED

- Copy and cut out one *Remainder Race Game Board* for each pair of players.
- Copy and cut out one of the *Remainder Race Game Markers* for each player.
- Collect a number cube, individual whiteboards and dry-erase markers, and scratch paper for each pair of players.

Number of Players

2 Players

Materials

- *Remainder Race Game Board* (pages 65–66)
- *Remainder Race Game Markers* (page 67)
- number cube or dice
- individual whiteboards and dry-erase markers
- scratch paper

Game Directions

1. Distribute materials to players.
2. Players take turns rolling a number cube. The player who rolls the lower number is Player 1.
3. Player 1 rolls the number cube and moves forward that number of spaces on the game board.

© Shell Education

Remainder Race *(cont.)*

4. Using the number rolled as the divisor, Player 1 divides it into the number on the space. The player can use individual whiteboards and markers, or paper and pencil, to help solve the problem.

5. Player 2 checks Player 1's work. If the quotient is correct, Player 1 remains on the space. If the quotient is incorrect, Player 1 must return to the space where he or she started.

6. The first player to reach the "Finish Line" wins the race.

© Shell Education

Remainder Race

Game Board

Directions: Cut out the game board. Tape it to the game board on page 66.

7,392

6,528

7,293

Finish Line

5,400

1,835

9,783

2,874

REM

RA

tape here

5,338

633

1,284

8,843

94

© Shell Education

Remainder Race

Game Board *(cont.)*

Start	152	760	832

1,045

MAINDER

287

CE

3,347

8,623

5	1,156	756	371	255

© Shell Education

Remainder Race

Game Markers

Directions: Copy and cut out a game marker for each player. Make sure each player uses a different color to identify his or her marker.

What's My Name?

Domain

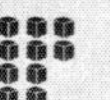

Number and Operations in Base Ten

Standard

Read and write multi-digit whole numbers using base-ten numerals, number names, and expanded form.

GET PREPARED!

- Copy and cut out one set of the *What's My Name? Number Cards* for each pair of players.
- Collect number cubes for each pair of players.

Number of Players

2 Players

Materials

- *What's My Name? Number Cards* (pages 70–75)
- number cubes

Game Directions

1. Distribute materials to players.
2. Players take turns rolling a number cube. The player who rolls the higher number is Player 1.
3. Player 1 shuffles the deck and deals six cards to each player. The rest of the cards are placed facedown in a draw pile between the two players.

 © Shell Education

What's My Name? *(cont.)*

4. Players check the cards in their hand for matching sets. A matching set has three cards that name the same number in numerals, words, and expanded form. If a player has a matching set before play begins, he or she can place that set to the side and draw three more cards.

5. Player 1 draws a card from the draw pile. If it matches any of the cards in his or her hand the player groups it with the set. The player must place one card in the discard pile.

6. Player 2 draws a card either from the draw pile or the discard pile. If it matches any of the cards in his or her hand, the player groups it with the set. The player must place one card in the discard pile.

7. The first player to get two sets of three cards wins the game.

8. If necessary, the discard pile is reshuffled and placed facedown.

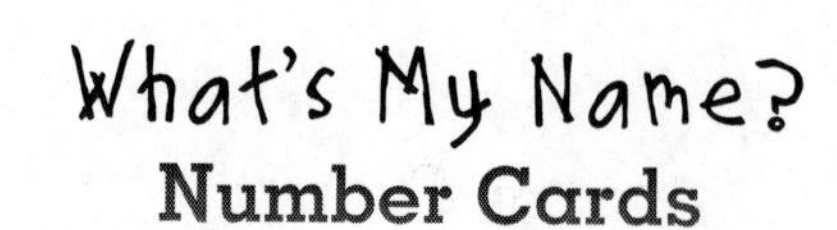

What's My Name?

Number Cards

Directions: Copy and cut out one set of cards for each pair of players.

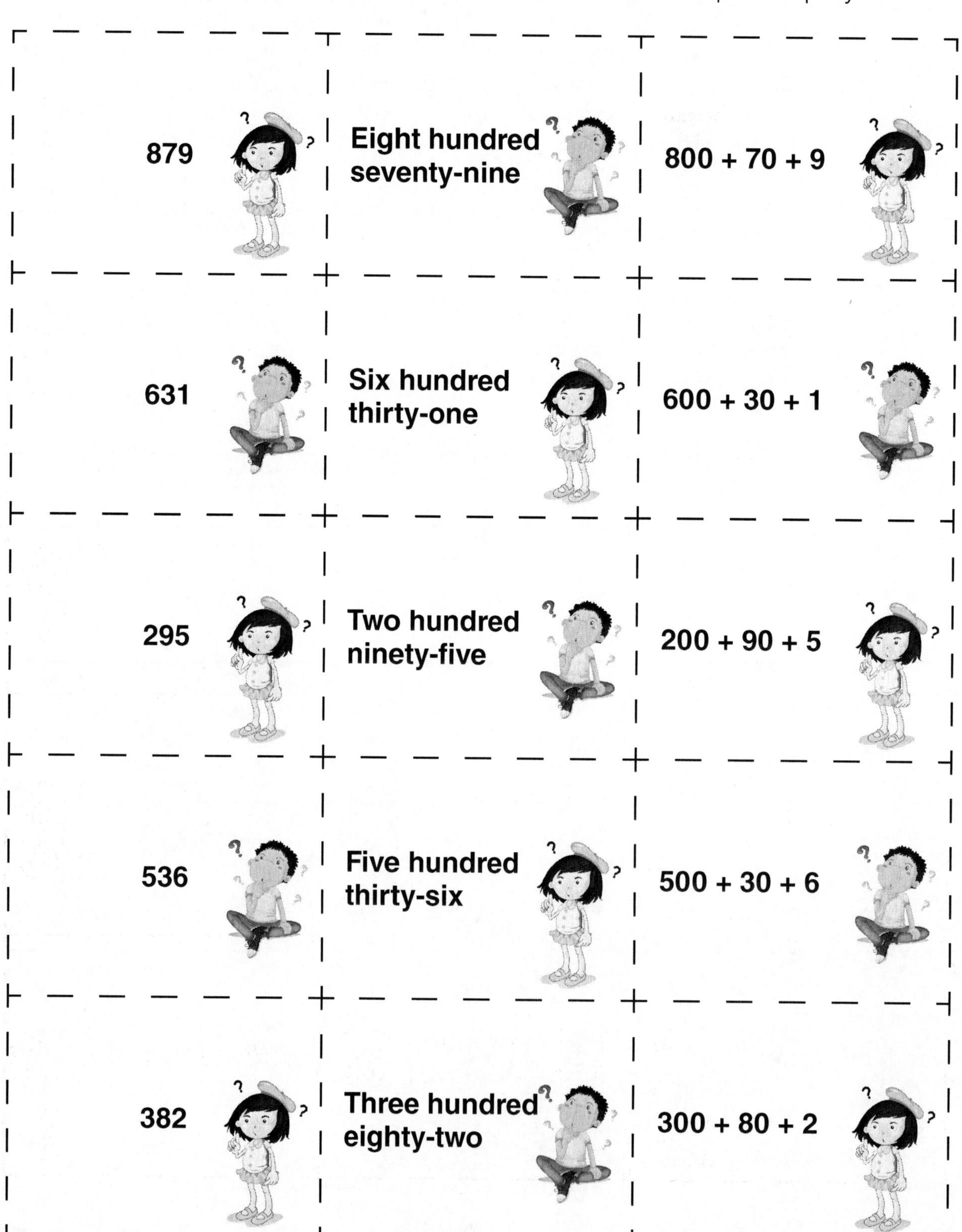

879	Eight hundred seventy-nine	800 + 70 + 9
631	Six hundred thirty-one	600 + 30 + 1
295	Two hundred ninety-five	200 + 90 + 5
536	Five hundred thirty-six	500 + 30 + 6
382	Three hundred eighty-two	300 + 80 + 2

© Shell Education

What's My Name?

Number Cards *(cont.)*

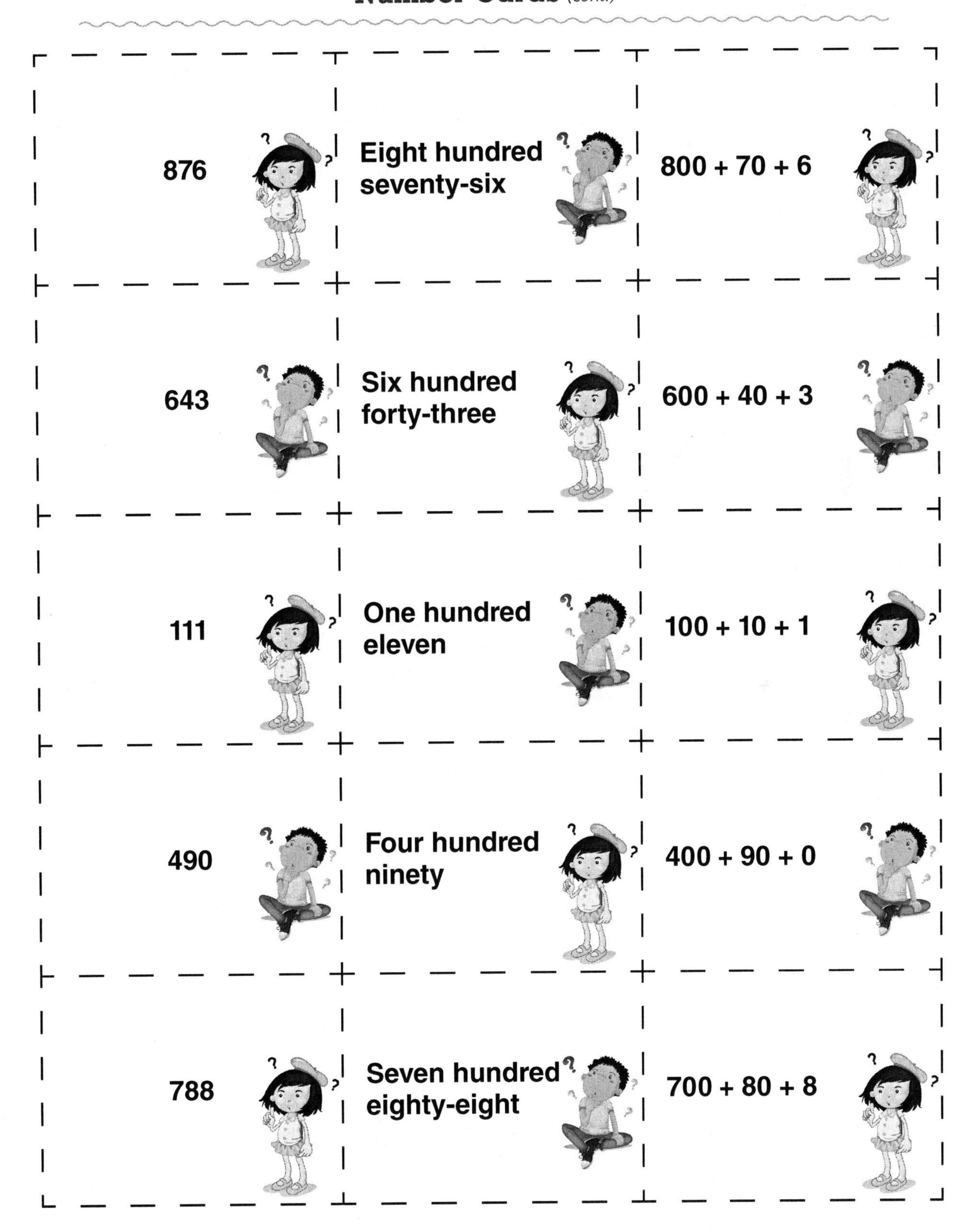

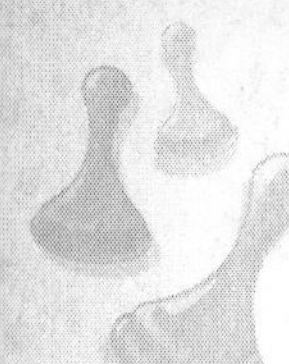

What's My Name?
Number Cards *(cont.)*

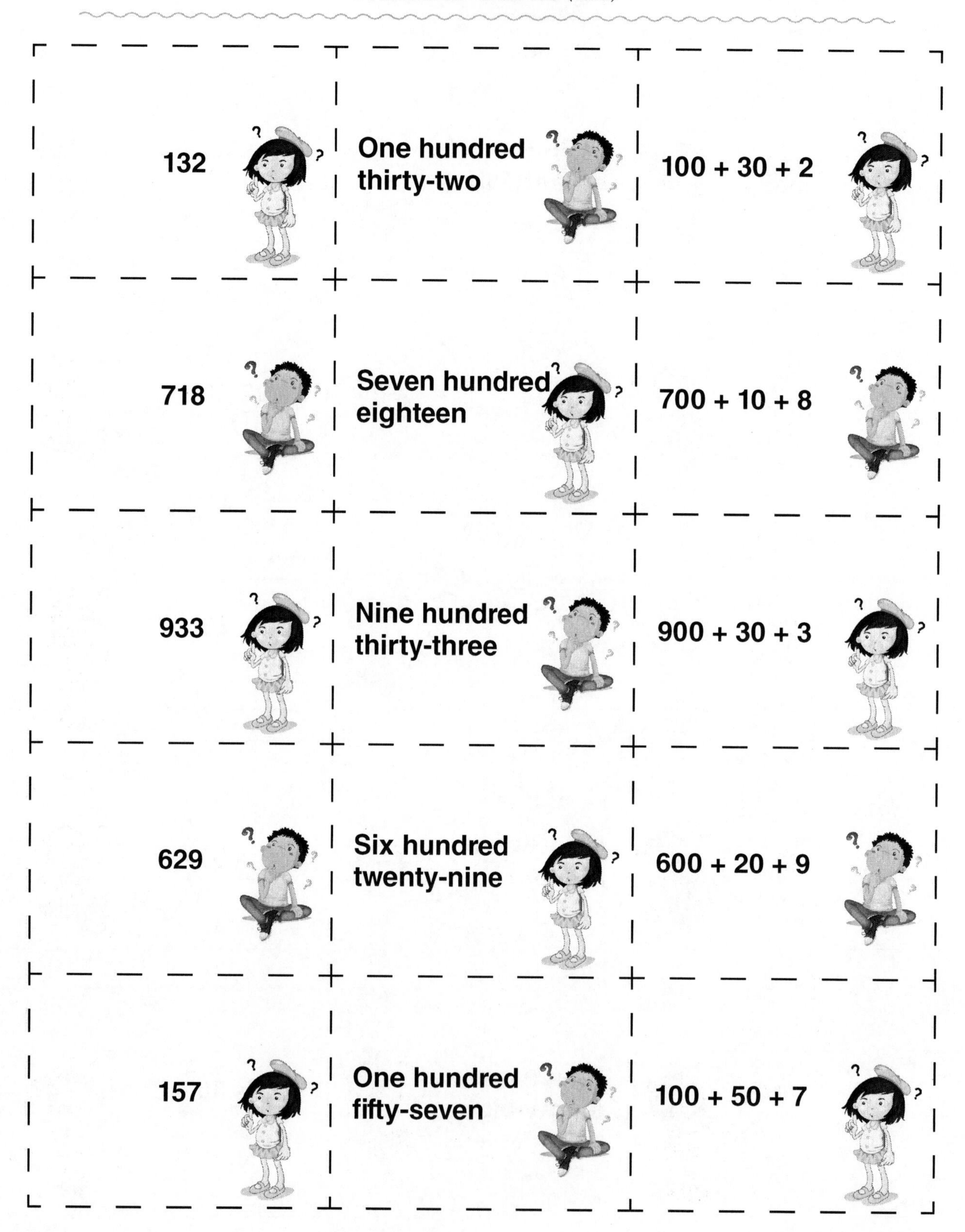

 © Shell Education

What's My Name?

Number Cards *(cont.)*

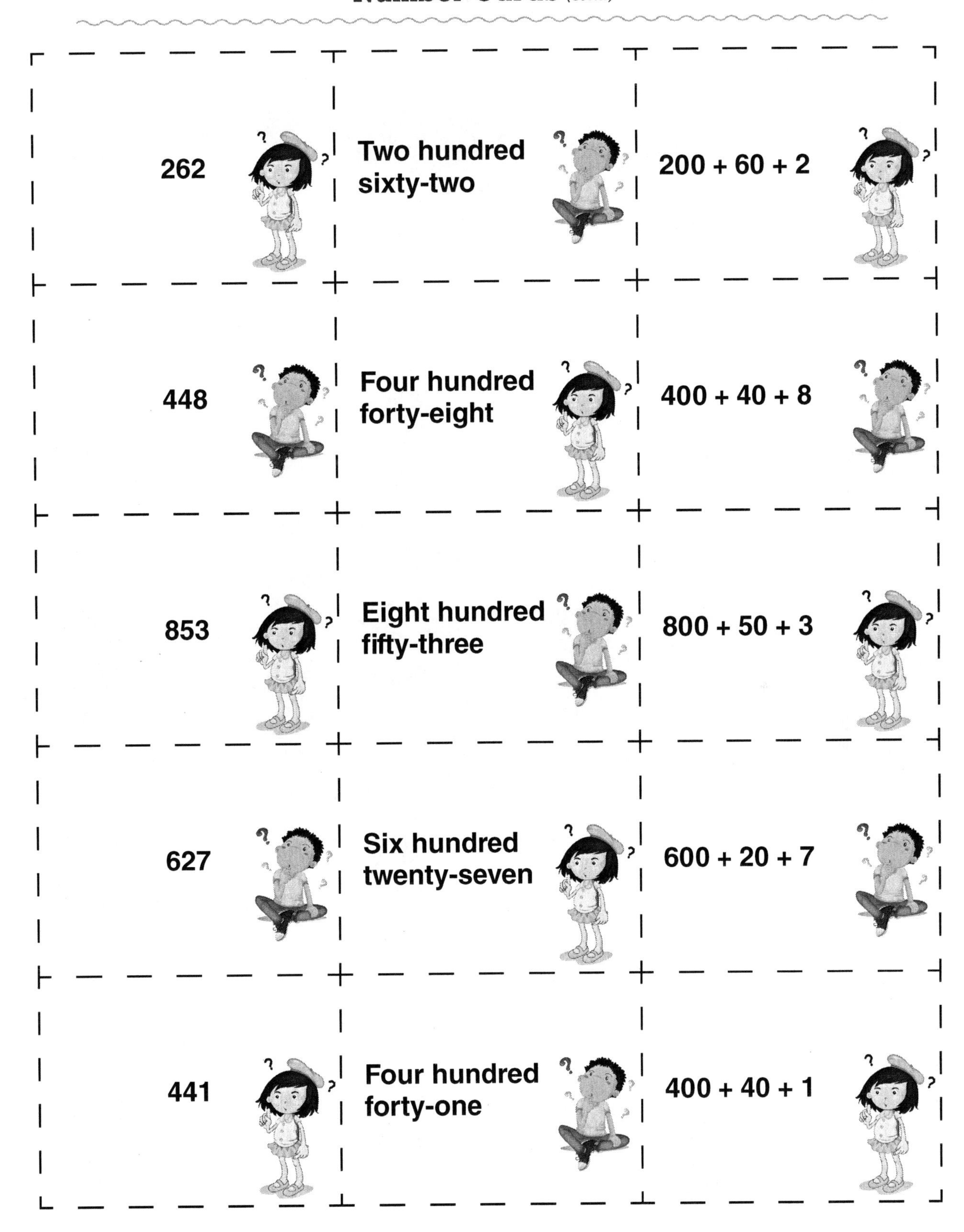

262	Two hundred sixty-two	200 + 60 + 2
448	Four hundred forty-eight	400 + 40 + 8
853	Eight hundred fifty-three	800 + 50 + 3
627	Six hundred twenty-seven	600 + 20 + 7
441	Four hundred forty-one	400 + 40 + 1

Number Cards *(cont.)*

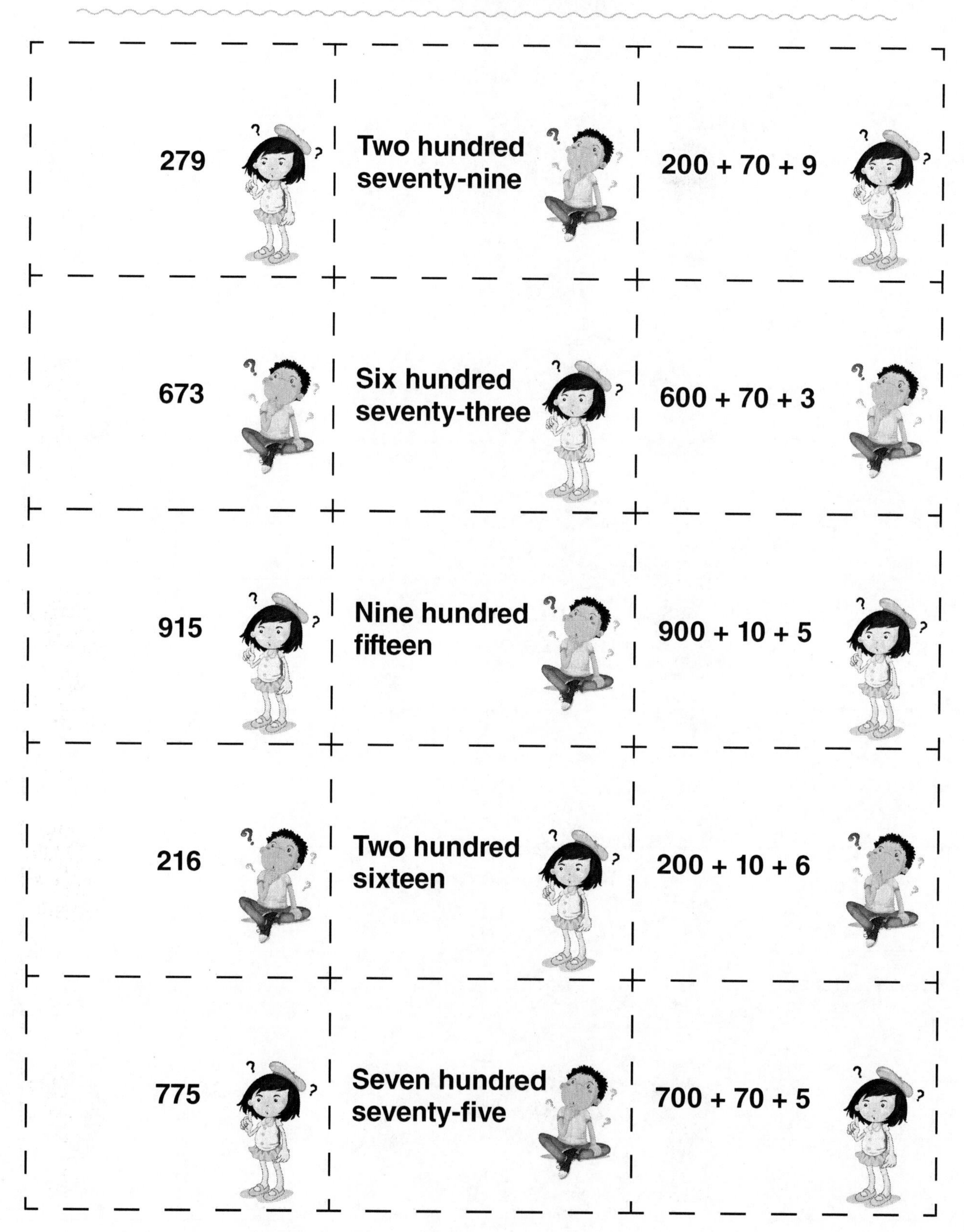

 © Shell Education

What's My Name?

Number Cards *(cont.)*

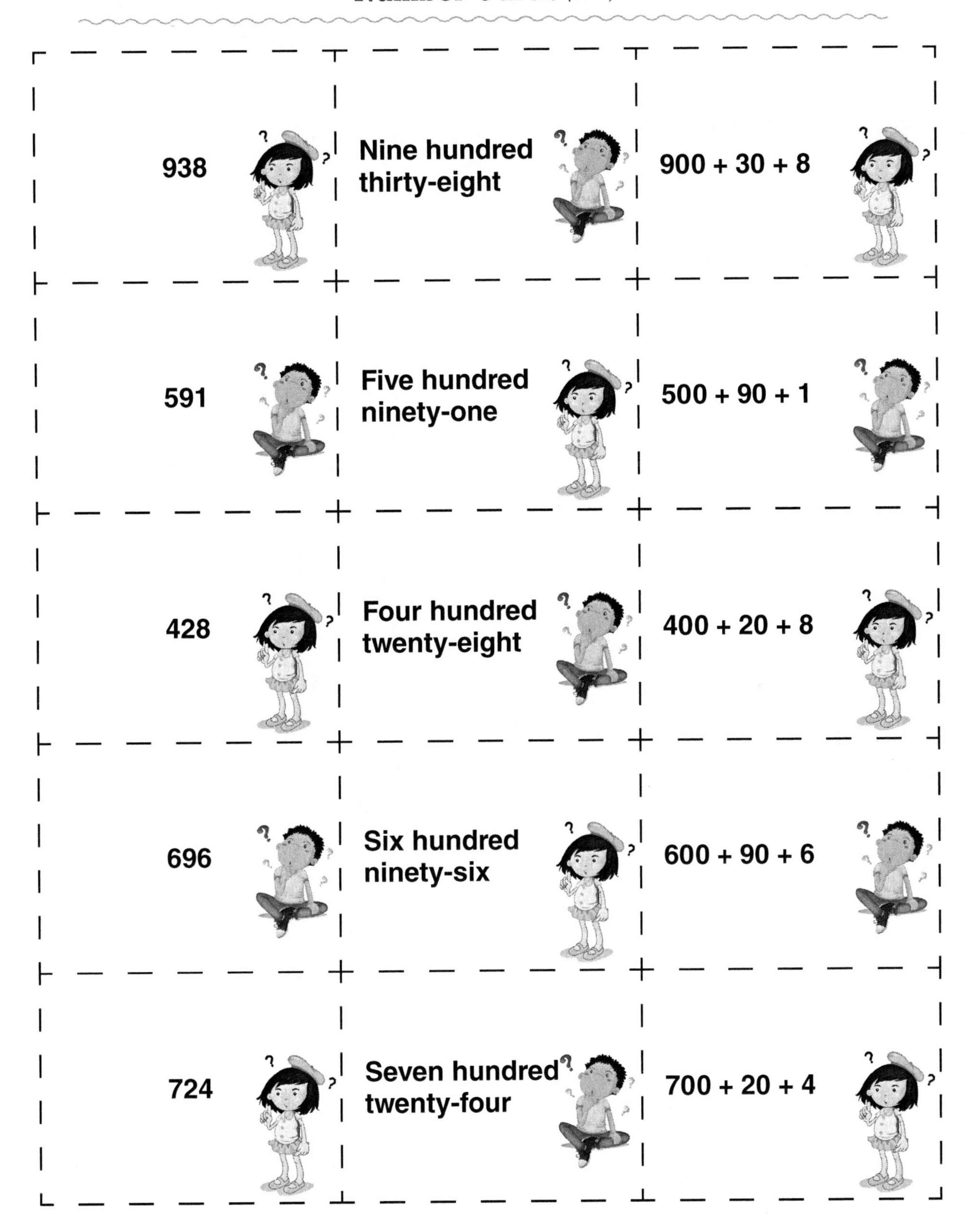

© Shell Education

Fishing for Fractions

Domain 

Number and Operations—Fractions

Standards 

Understand a fraction a/b with a > 1 as a sum of fractions 1/b.

Decompose a fraction into a sum of fractions with the same denominator in more than one way, recording each decomposition by an equation.

Number of Players 

2 Players

Materials 

- *Fishing for Fractions Cards* (pages 78–80)
- number cubes

GET PREPARED!

- Copy and cut out one set of the *Fishing for Fractions Cards* for each pair of players.
- Collect number cubes for each pair of players.

Game Directions

1. Distribute materials to players. The cards are pairs that name the same fraction, similar to *Go Fish!*
2. Players take turns rolling a number cube. The player who rolls the higher number is Player 1.
3. Player 1 deals five cards facedown to each player. Players check the cards in their hands for matching pairs.

Fishing for Fractions *(cont.)*

4. Player 1 begins by "fishing" for a fraction card that matches a card in his or her hand. For example, Player 1 says, "I have three eighths. Do you have three eighths?"

5. If Player 2 has the matching card, he or she must give it to Player 1. For example, if Player 2 has "$\frac{1}{8}+\frac{1}{8}+\frac{1}{8}$," he or she gives it to Player 1.

6. If Player 1 gets a pair, he or she places it on the desk and asks Player 2 for another card. This continues as long as Player 2 has the requested cards. Both players will draw cards to ensure that they always have five cards in hand.

7. When Player 1 asks for a card that Player 2 does not have, Player 2 says, "Go Fish!" That means Player 1 must draw a card from the deck and add it to his or her hand. If the card from the deck happens to be the one Player 1 was "fishing" for, he or she can place the pair on the table.

8. Player 2 repeats steps 4 to 7.

9. The game continues until all fractions are matched. The player with more pairs of cards wins the game.

Fishing for Fractions

Cards

Directions: Copy and cut out the cards for each group of players.

$\frac{2}{3}$	$\frac{1}{3} + \frac{1}{3}$	$\frac{3}{4}$
$\frac{1}{4} + \frac{1}{4} + \frac{1}{4}$	$\frac{4}{5}$	$\frac{1}{5} + \frac{3}{5}$
$\frac{2}{5}$	$\frac{1}{5} + \frac{1}{5}$	$\frac{5}{6}$
$\frac{2}{6} + \frac{3}{6}$	$\frac{3}{6}$	$\frac{1}{6} + \frac{1}{6} + \frac{1}{6}$

© Shell Education

Fishing for Fractions

Cards *(cont.)*

$\frac{7}{8}$	$\frac{1}{8} + \frac{2}{8} + \frac{4}{8}$	$\frac{6}{8}$
$\frac{1}{8} + \frac{2}{8} + \frac{3}{8}$	$\frac{5}{8}$	$\frac{1}{8} + \frac{3}{8} + \frac{1}{8}$
$\frac{8}{10}$	$\frac{1}{10} + \frac{7}{10}$	$\frac{7}{10}$
$\frac{3}{10} + \frac{4}{10}$	$\frac{5}{10}$	$\frac{3}{10} + \frac{2}{10}$

Fishing for Fractions

Cards *(cont.)*

$\frac{3}{10}$	$\frac{2}{10} + \frac{1}{10}$	$\frac{5}{12}$
$\frac{1}{12} + \frac{4}{12}$	$\frac{7}{12}$	$\frac{1}{12} + \frac{6}{12}$
$\frac{11}{12}$	$\frac{7}{12} + \frac{4}{12}$	$\frac{3}{5}$
$\frac{1}{5} + \frac{1}{5} + \frac{1}{5}$	$\frac{9}{10}$	$\frac{3}{10} + \frac{6}{10}$

© Shell Education

Dueling Decimals

Domain

Number and Operations—Fractions

Standard

Compare two decimals to hundredths by reasoning about their size. Recognize that comparisons are valid only when the two decimals refer to the same whole. Record the results of comparisons with the symbols >, =, or <, and justify the conclusions (e.g., by using a visual model).

GET PREPARED

- Copy the *Dueling Decimals Recording Sheets* for each player.
- Copy and cut out the *Dueling Decimals Spinner* for each pair of players.
- Collect one paperclip and one sharpened pencil for each pair of players to use for their spinners.
- Collect a number cube for each pair of players.

Number of Players

2 Players

Materials

- *Dueling Decimals Spinner* (page 83)
- *Dueling Decimals Recording Sheet* (page 84)
- paperclips and sharpened pencils
- number cubes

Game Directions

1. Distribute materials to players.
2. Players take turns rolling a number cube. The player who rolls the lower number is Player 1.
3. Player 1 flicks the paperclip on the spinner and records the digit it lands on in either the tenths or hundredths place on his or her recording sheet. Player 1 spins a second time to select the digit to record in the remaining space on his or her recording sheet.

Dueling Decimals *(cont.)*

4. Player 2 spins and records the digit selected in the tenths or hundredths place on his or her recording sheet. The player spins a second time to select the digit to write in the remaining space on his or her recording sheet.

5. If the number they made makes their statement true, players earn one point. If the number makes their statement false, they do not earn a point.

		Score
0.25	< 0.35	**1**
0.61	> 0.65	

6. Players play 10 rounds. The player with the higher score at the end of the last round wins the game.

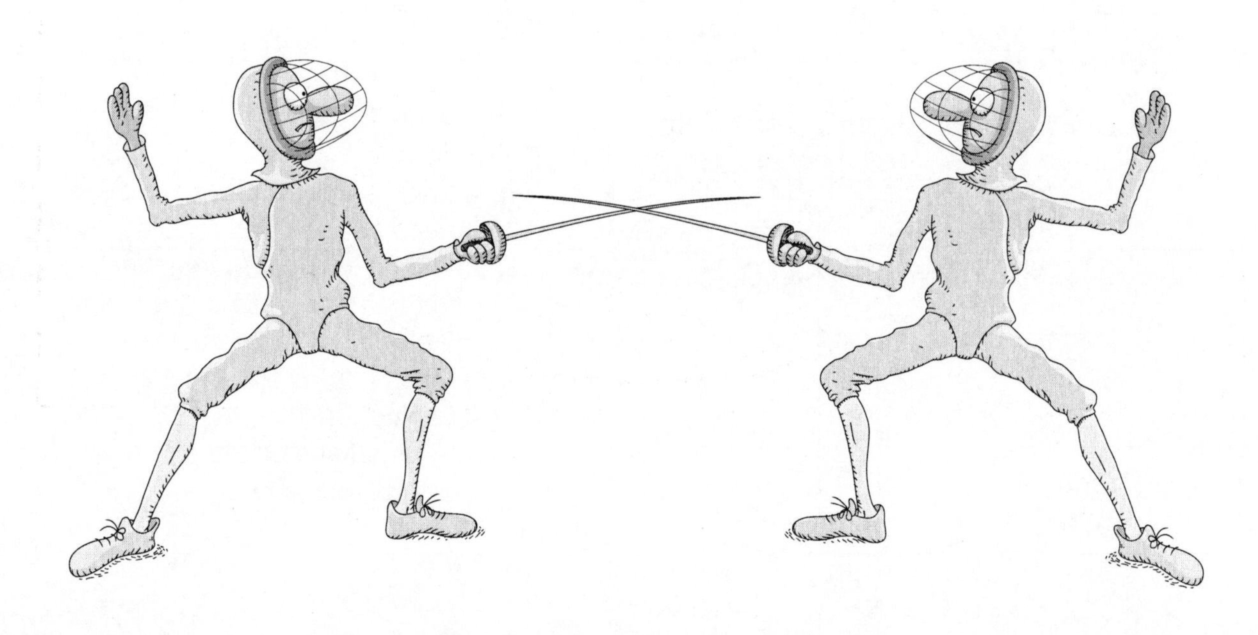

© Shell Education

DUELING DECIMALS

Spinner

Directions: Copy and cut out a spinner for each pair of players. For directions on how to assemble this spinner, see page 10.

Name: ____________________________ Date: ________________

DUELING DECIMALS

Recording Sheet

Directions: Record the digits spun on this sheet.

		Score
0. ______ ______	< 0.35	
0. ______ ______	> 0.65	
0. ______ ______	> 0.43	
0. ______ ______	< 0.31	
0. ______ ______	> 0.18	
0. ______ ______	> 0.21	
0. ______ ______	< 0.76	
0. ______ ______	> 0.43	
0. ______ ______	< 0.31	
0. ______ ______	< 0.87	

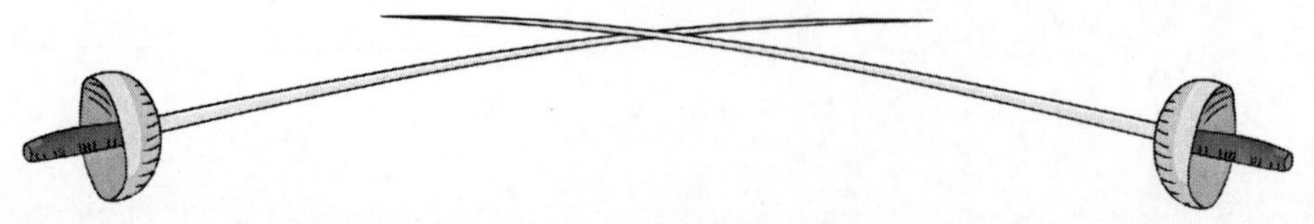

© Shell Education

Fraction Flip

Domain

Number and Operations—Fractions

Standards

Compare two fractions with different numerators and different denominators (e.g., by creating common denominators or numerators, or by comparing to a benchmark fraction such as $\frac{1}{2}$).

Recognize that comparisons are valid only when the two fractions refer to the same whole.

Record the results of comparisons with symbols >, =, or <, and justify the conclusions, (e.g., by using a visual fraction model).

Number of Players

2 to 3 Players

Materials

- *Fraction Flip Cards* (pages 87–89)
- *Fraction Flip Strips* (page 90)
- number cubes

GET PREPARED!

- Copy and cut out one set of the *Fraction Flip Cards* and the *Fraction Flip Strips* for each group of players.
- Collect a number cube for each group of players.

Game Directions

1. Distribute materials to players.
2. Players take turns rolling a number cube. The player who rolls the highest number is Player 1. The player who rolls the second highest number is Player 2, and the player who rolls the lowest number is Player 3.
3. Player 1 shuffles the cards and places them facedown in the center of the playing area.

Fraction Flip *(cont.)*

4. Player 1 draws a card and flips it faceup.
5. Players 2 and 3 repeat step 4. The player with the largest fraction wins the round. The winner collects all cards.
6. Play continues until all cards have been played. The player with the most cards at the end of the game is the winner.
7. In case players cannot determine which fraction is greater, they use the *Fraction Flip Strips* to compare fractions.
8. A second version of the game can be played where the player with the lowest fraction wins each round.

© Shell Education

Fraction Flip

Cards

Directions: Copy and cut out the cards for each group of players.

$\frac{1}{2}$	$\frac{3}{4}$	$\frac{4}{5}$
$\frac{1}{3}$	$\frac{1}{5}$	$\frac{1}{6}$
$\frac{2}{3}$	$\frac{2}{5}$	$\frac{3}{6}$
$\frac{1}{4}$	$\frac{3}{5}$	$\frac{4}{6}$

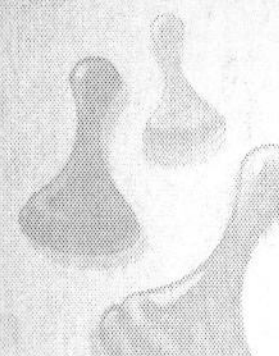

Fraction Flip

Cards *(cont.)*

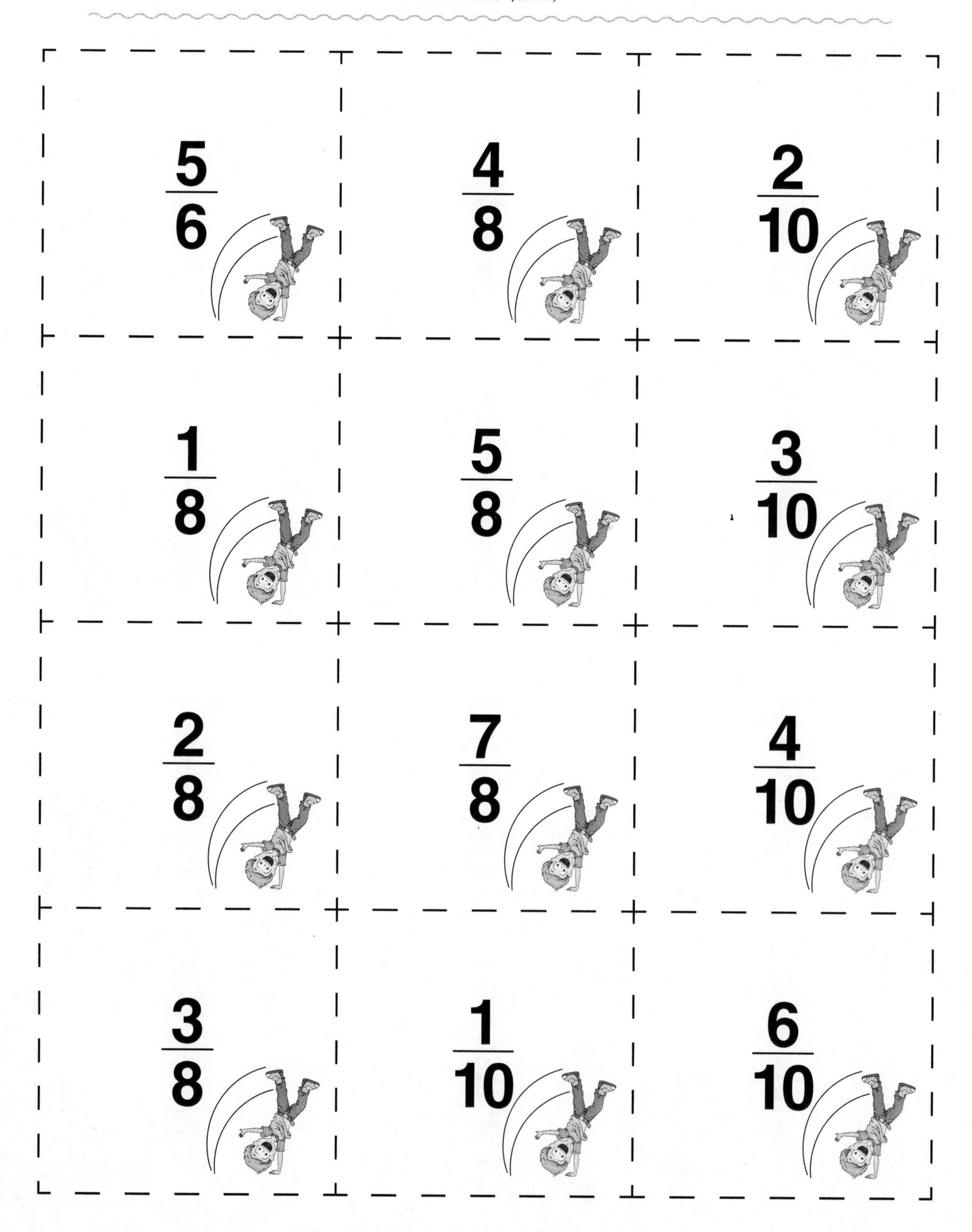

© Shell Education

Cards *(cont.)*

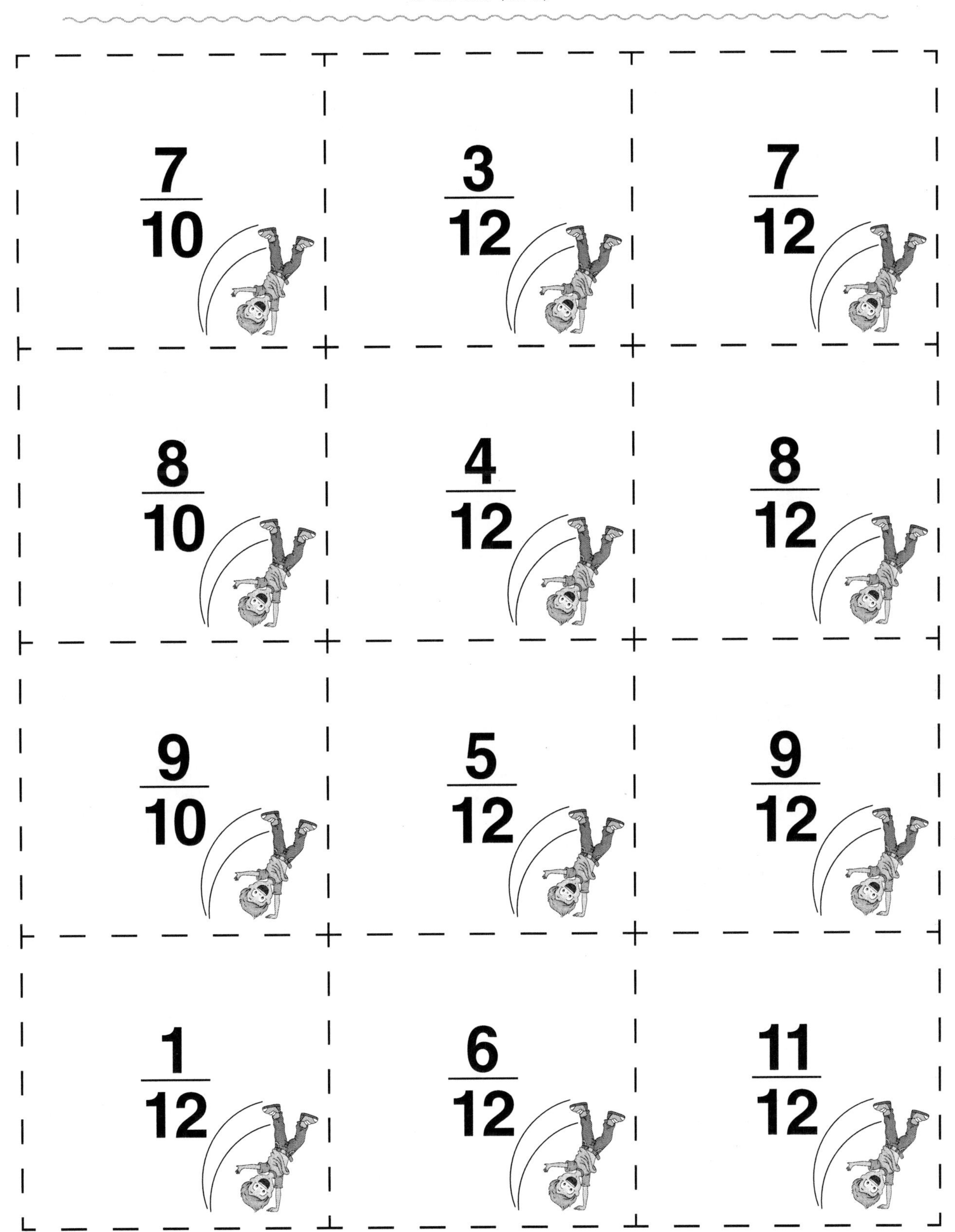

Fraction Flip

Strips

Directions: Use the fraction strip to determine which fraction is greater.

$\frac{1}{2}$ | $\frac{1}{2}$

$\frac{1}{3}$ | $\frac{1}{3}$ | $\frac{1}{3}$

$\frac{1}{4}$ | $\frac{1}{4}$ | $\frac{1}{4}$ | $\frac{1}{4}$

$\frac{1}{5}$ | $\frac{1}{5}$ | $\frac{1}{5}$ | $\frac{1}{5}$ | $\frac{1}{5}$

$\frac{1}{6}$ | $\frac{1}{6}$ | $\frac{1}{6}$ | $\frac{1}{6}$ | $\frac{1}{6}$ | $\frac{1}{6}$

$\frac{1}{8}$ | $\frac{1}{8}$ | $\frac{1}{8}$ | $\frac{1}{8}$ | $\frac{1}{8}$ | $\frac{1}{8}$ | $\frac{1}{8}$ | $\frac{1}{8}$

$\frac{1}{10}$ | $\frac{1}{10}$ | $\frac{1}{10}$ | $\frac{1}{10}$ | $\frac{1}{10}$ | $\frac{1}{10}$ | $\frac{1}{10}$ | $\frac{1}{10}$ | $\frac{1}{10}$ | $\frac{1}{10}$

$\frac{1}{12}$ | $\frac{1}{12}$ | $\frac{1}{12}$ | $\frac{1}{12}$ | $\frac{1}{12}$ | $\frac{1}{12}$ | $\frac{1}{12}$ | $\frac{1}{12}$ | $\frac{1}{12}$ | $\frac{1}{12}$ | $\frac{1}{12}$ | $\frac{1}{12}$

© Shell Education

Fraction Decimal Match

Domain

Number and Operations—Fractions

Standard

Use decimal notation for fractions with denominators 10 or 100.

GET PREPARED!

- Copy and cut out one set of the *Fraction Decimal Match Cards* for each pair of players.
- Collect a number cube for each pair of players.

Number of Players

2 Players

Materials

- *Fraction Decimal Match Cards* (page 93–96)
- number cubes

Game Directions

1. Distribute materials to players.
2. Players take turns rolling a number cube. The player who rolls the lower number is Player 1.
3. Players arrange *Fraction Decimal Match Cards* facedown in a 4 × 8 array.

Fraction Decimal Match *(cont.)*

4. Player 1 turns over two cards in order to find a matching pair. A matching pair is a fraction and a decimal that represent the same value.
5. If the cards match, Player 1 keeps the pair and goes again.
6. If the cards do not match, Player 1 returns them to the same place in the array.
7. Player 2 repeats steps 4 to 6.
8. Play continues until no cards remain. The player with more pairs at the end of the game is the winner.

© Shell Education

Fraction Decimal Match
Cards

Directions: Copy and cut out cards for each pair of players.

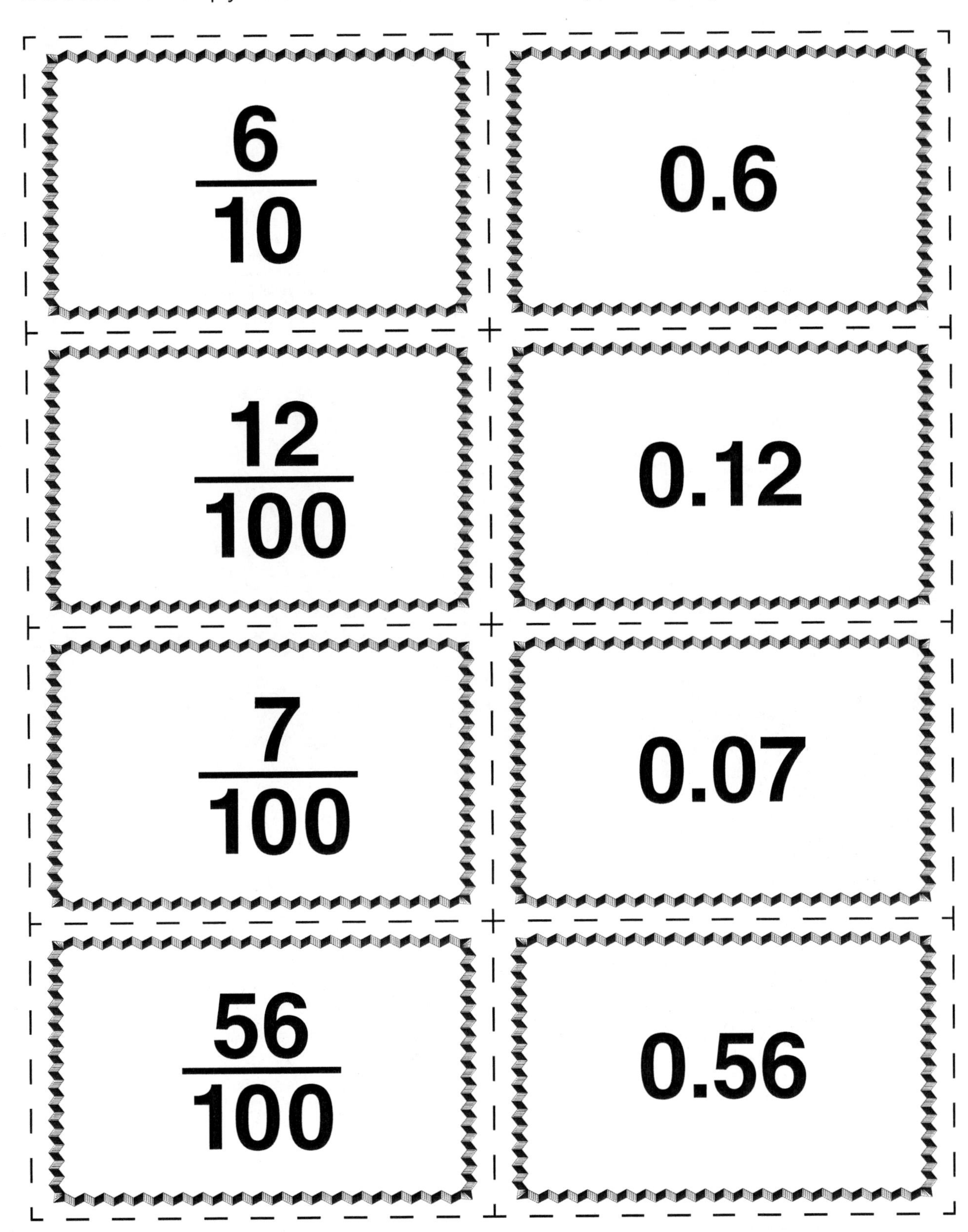

Cards *(cont.)*

$\frac{8}{10}$	0.8
$\frac{73}{100}$	0.73
$\frac{3}{100}$	0.03
$\frac{27}{100}$	0.27

© Shell Education

Fraction Decimal Match

Cards *(cont.)*

$\frac{38}{100}$	0.38
$\frac{8}{100}$	0.08
$\frac{70}{100}$	0.7
$\frac{4}{10}$	0.4

Fraction Decimal Match

Cards *(cont.)*

$\frac{4}{100}$	0.04
$\frac{92}{100}$	0.92
$\frac{9}{100}$	0.09
$\frac{5}{10}$	0.5

© Shell Education

Equivalent Fractions Bingo

Domain

Number and Operations—Fractions

Standard

Explain why a fraction $\frac{a}{b}$ is equivalent to a fraction $\frac{(n \times a)}{(n \times b)}$ by using visual fraction models, with attention to how the number and size of the parts differ even though the two fractions themselves are the same size. Use this principle to recognize and generate equivalent fractions.

Number of Players

Whole class

Materials

- *Equivalent Fractions Bingo Board* (page 99)
- *Equivalent Fractions Cards* (pages 100–101)
- paper bag
- bingo markers (e.g., colored counters)

GET PREPARED!

- Copy and cut out one set of the *Equivalent Fractions Cards* for the class and place them in a paper bag.
- Copy and cut out the *Equivalent Fractions Bingo Board* for each player.
- Collect small items for players to use as bingo markers.
- On the board, write these fractions: $\frac{1}{2}$, $\frac{1}{3}$, $\frac{2}{3}$, $\frac{1}{4}$, $\frac{3}{4}$, $\frac{1}{5}$, $\frac{2}{5}$, $\frac{3}{5}$, $\frac{4}{5}$, $\frac{1}{6}$, and $\frac{5}{6}$.

Game Directions

1. Distribute materials to players.
2. Each player receives a blank *Equivalent Fractions Bingo Board* and bingo markers. Players write any nine of the fractions displayed on the whiteboard onto their bingo boards in any order.

Equivalent Fractions Bingo *(cont.)*

3. The teacher draws a card from the bag and reads the fraction aloud.

4. Players look for an equivalent fraction on their bingo boards. If the fraction is on their board, players place a marker on the space.

5. The teacher continues to draw cards from the bag, and players mark the equivalent fractions. The first player to mark three spaces in a row—horizontally, vertically, or diagonally—calls "Bingo!" If the spots recorded are correct, he or she wins the game.

© Shell Education

Equivalent Fractions

Bingo Board

Directions: Copy and cut out one bingo board for each player.

Equivalent Fractions Bingo

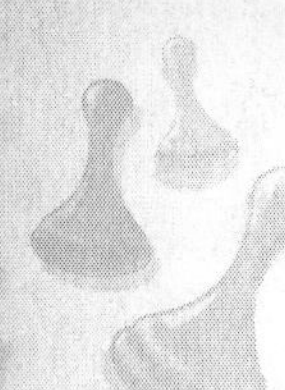

Equivalent Fractions

Cards

Directions: Copy and cut out the cards for the class.

$\frac{9}{12}$	$\frac{5}{15}$	$\frac{3}{18}$
$\frac{5}{10}$	$\frac{6}{12}$	$\frac{2}{6}$
$\frac{4}{12}$	$\frac{2}{8}$	$\frac{3}{12}$

© Shell Education

Equivalent Fractions

Cards *(cont.)*

$\frac{2}{12}$	$\frac{4}{6}$	$\frac{8}{12}$
$\frac{6}{8}$	$\frac{2}{10}$	$\frac{10}{12}$
$\frac{6}{10}$	$\frac{8}{10}$	$\frac{4}{10}$

© Shell Education

Fraction Action Race

Domain

Number and Operations—Fractions

Standards

Compare two fractions with different numerators and different denominators, e.g., by creating common denominators or numerators, or by comparing to a benchmark fraction such as $\frac{1}{2}$.

Recognize that comparisons are valid only when the two fractions refer to the same whole.

Record the results of comparisons with symbols >, =, or <, and justify the conclusions, e.g., by using a visual fraction model.

Number of Players

2 Players

Materials

- *Fraction Action Race Game Board* (pages 104–105)
- *Fraction Action Race Fraction Strips* (page 106)
- number cubes
- game markers

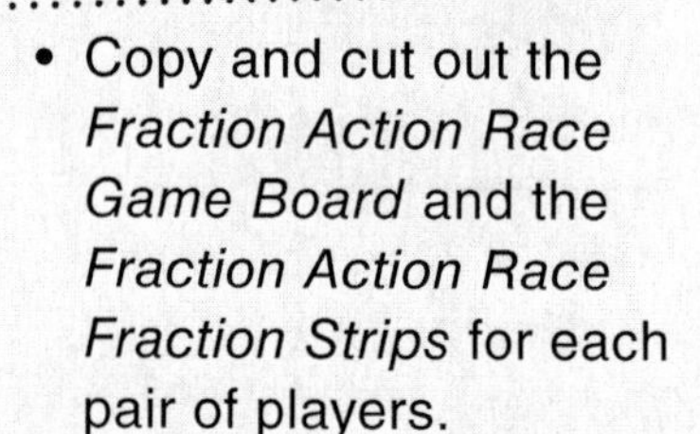

- Copy and cut out the *Fraction Action Race Game Board* and the *Fraction Action Race Fraction Strips* for each pair of players.
- Collect two number cubes and two game markers for each pair of players.

Game Directions

1. Distribute materials to players.
2. Players take turns rolling a number cube. The player who rolls the lower number is Player 1.
3. Player 1 rolls two number cubes and creates a proper fraction. The first roll is the numerator, and the second roll is the denominator If the player rolls "doubles," he or she loses that turn since a proper fraction cannot be formed.

© Shell Education

Fraction Action Race *(cont.)*

4. Player 1 compares the new fraction to the fraction in the first square of the game board. If the new fraction is greater, the player moves his or her game marker to that square. If the fraction is less than the number on the square, the player does not move his or her game piece. Players may use the *Fraction Action Race Fraction Strips* to make comparisons. If the new fraction is the same as the fraction on the board, the player chooses one number cube to roll again to create a different fraction.

5. Player 2 rolls the number cubes, forms a proper fraction, and compares the fraction to the fraction in the first space on the board.

6. Play continues until one player reaches the finish line. The first player to reach the "Finish" space wins the game.

Fraction Action Race

Game Board

Directions: Copy and cut out the game board. Tape it to the game board on page 105.

START	$\frac{1}{2}$	$\frac{1}{3}$
FINISH $\frac{3}{4}$	Fracti Actio Rac	
$\frac{2}{3}$	$\frac{1}{6}$	$\frac{1}{5}$

 © Shell Education

Fraction Action Race

Game Board *(cont.)*

Fraction Action Race

Fraction Strips

Directions: Use the fraction strip to determine which fraction is greater.

$\frac{1}{2}$	$\frac{1}{2}$								
$\frac{1}{3}$	$\frac{1}{3}$	$\frac{1}{3}$							
$\frac{1}{4}$	$\frac{1}{4}$	$\frac{1}{4}$	$\frac{1}{4}$						
$\frac{1}{5}$	$\frac{1}{5}$	$\frac{1}{5}$	$\frac{1}{5}$	$\frac{1}{5}$					
$\frac{1}{6}$	$\frac{1}{6}$	$\frac{1}{6}$	$\frac{1}{6}$	$\frac{1}{6}$	$\frac{1}{6}$				
$\frac{1}{8}$	$\frac{1}{8}$	$\frac{1}{8}$	$\frac{1}{8}$	$\frac{1}{8}$	$\frac{1}{8}$	$\frac{1}{8}$	$\frac{1}{8}$		
$\frac{1}{10}$	$\frac{1}{10}$	$\frac{1}{10}$	$\frac{1}{10}$	$\frac{1}{10}$	$\frac{1}{10}$	$\frac{1}{10}$	$\frac{1}{10}$	$\frac{1}{10}$	$\frac{1}{10}$

 © Shell Education

Improper Fraction Bingo

Domain

Number and Operations—Fractions

Standard

Add and subtract mixed numbers with like denominators, e.g., by replacing each mixed number with an equivalent fraction, and/or by using properties of operations and the relationship between addition and subtraction.

Number of Players

Groups of 3 or 4

Materials

- *Improper Fraction Bingo Board* (page 109)
- *Improper Fraction Bingo Spinner* (page 110)
- *Improper Fraction Recording Sheet* (page 111)
- paperclips and sharpened pencils
- bingo markers (e.g., bingo chips, colored counters, or coins)
- scratch paper

GET PREPARED!

- Copy and cut out the *Improper Fraction Bingo Board* and the *Improper Fraction Recording Sheet* for each player.
- Copy and cut out the *Improper Fraction Bingo Spinner* for each group of players.
- Collect one paperclip and one sharpened pencil for each group to use with their spinners.
- On the board, write the following improper fractons: $\frac{14}{6}$, $\frac{15}{6}$, $\frac{16}{6}$, $\frac{17}{6}$, $\frac{18}{6}$, $\frac{19}{6}$, $\frac{20}{6}$, $\frac{21}{6}$, $\frac{22}{6}$, $\frac{23}{6}$, $\frac{24}{6}$

Game Directions

1. Distribute materials to players.
2. Players copy any nine of the improper fractions from the list on the board onto their bingo boards in any order.

Improper Fraction Bingo *(cont.)*

3. Each player spins the spinner twice. Players write both mixed numbers on the *Improper Fraction Recording Sheet* and add them to determine the improper fraction.

4. Players look for the matching improper fraction on their bingo boards. If the fraction is on the board, players place a marker on the space.

5. The first player in the group to mark three spaces in a row—horizontally, vertically, or diagonally—calls "Bingo!" He or she wins the game.

Improper Fraction Bingo

$\frac{14}{6}$	$\frac{15}{6}$	$\frac{16}{6}$
$\frac{17}{6}$	$\frac{18}{6}$	$\frac{19}{6}$
$\frac{20}{6}$	$\frac{21}{6}$	$\frac{22}{6}$

© Shell Education

Improper Fraction

Bingo Board

Directions: Copy and cut out a bingo board for each player.

Improper Fraction Bingo

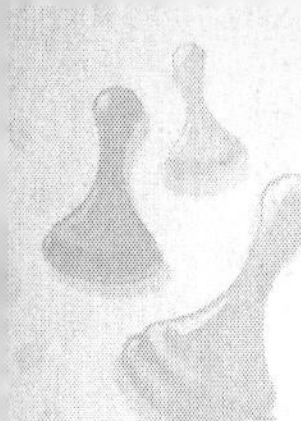

Improper Fraction

Bingo Spinner

Directions: Copy and cut out a spinner for each group of players. For steps on how to assemble this spinner, see page 10.

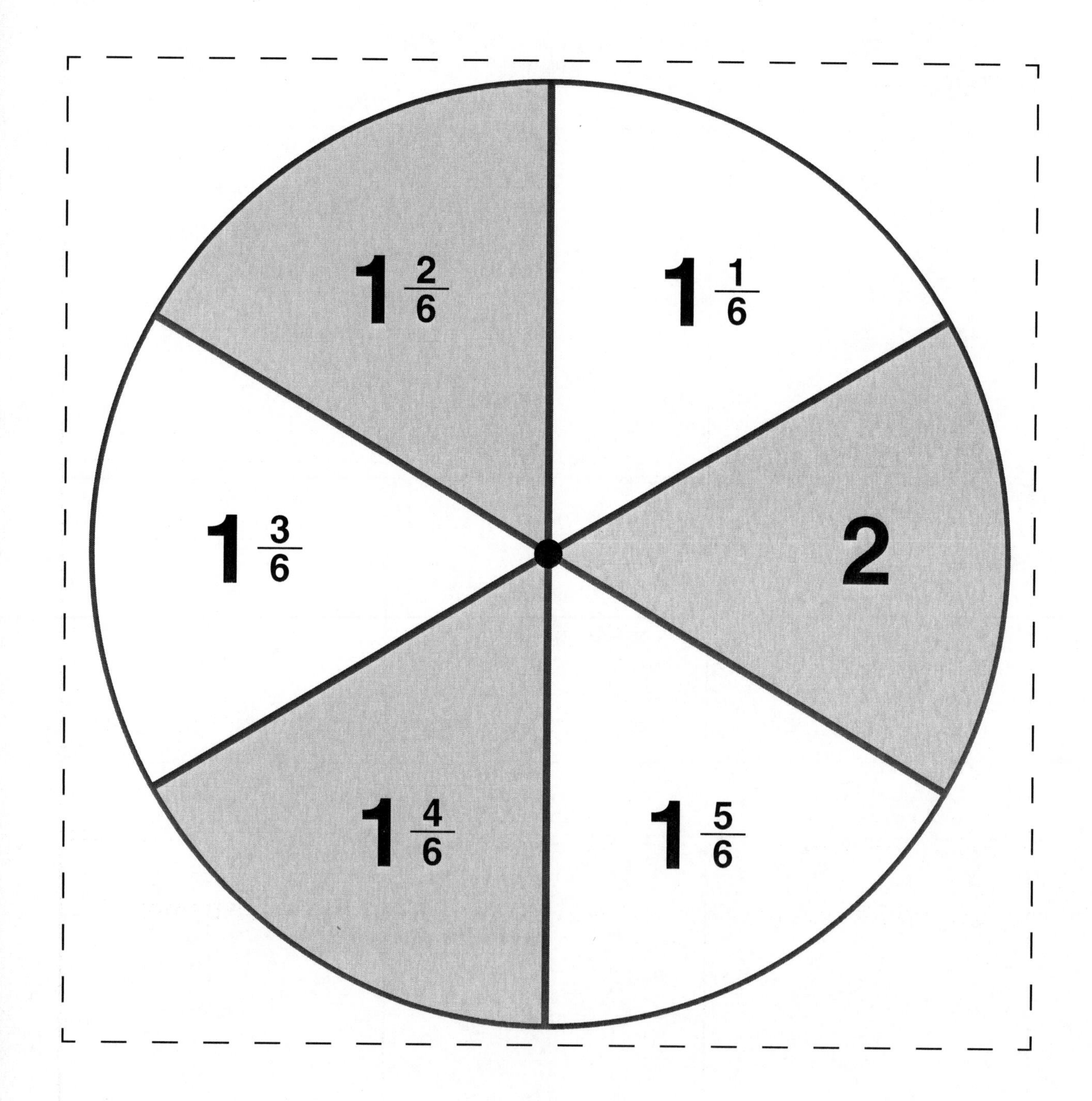

© Shell Education

Name: ______________________________ Date: ________________

Improper Fraction

Recording Sheet

Directions: Add the mixed numbers together to find the improper fraction.

Mixed Numbers: + Improper Fraction:	Mixed Numbers: + Improper Fraction:	Mixed Numbers: + Improper Fraction:
Mixed Numbers: + Improper Fraction:	Mixed Numbers: + Improper Fraction:	Mixed Numbers: + Improper Fraction:
Mixed Numbers: + Improper Fraction:	Mixed Numbers: + Improper Fraction:	Mixed Numbers: + Improper Fraction:

© Shell Education

Pathway Puzzles

Domain

Measurement and Data

Standard

Within a single system of measurement, express measurements in a larger unit in terms of a smaller unit.

Number of Players

2 Players

Materials

- *Pathway Puzzles Cards* (pages 114–117)
- *Pathway Puzzles Recording Sheet* (page 118)
- *Pathway Puzzles Game Board* (pages 119–120)
- *Pathway Puzzles Answer Key* (page 151)
- metric rulers
- number cubes

GET PREPARED

- Copy and cut out the *Pathway Puzzles Cards*, *Pathway Puzzles Game Board*, and *Pathway Puzzles Answer Key* for each pair of players.
- Copy and cut out two *Pathway Puzzles Recording Sheets* for each player.
- Collect one ruler and number cube for each pair of players.

Game Directions

1. Distribute materials to players.
2. Players take turns rolling a number cube. The player who rolls the higher number is Player 1.
3. Say, "On your mark. Get set. Go!" Player 1 draws one of the *Pathway Puzzles Cards* and measures each segment of the pathway in centimeters. He or she writes down the length above each segment on the *Pathway Puzzles Recording Sheet*.

© Shell Education

Pathway Puzzles *(cont.)*

4. Player 1 then adds the segments and converts the sum to millimeters.

5. Player 2 checks the answer using the *Pathway Puzzles Answer Key*. If the answer is correct, the card is placed in the discard pile. Player 1 rolls a number cube and moves that many spaces on the *Pathway Puzzles Game Board*. If Player 1 is incorrect, he or she does not move and the card is placed in the discard pile.

6. Player 2 repeats steps 3, 4, and 5.

7. Play continues until a player reaches the "Finish" space on the game board.

Pathway Puzzles

Cards

Directions: Copy and cut out the cards for each pair of players.

1

2

3

4

© Shell Education

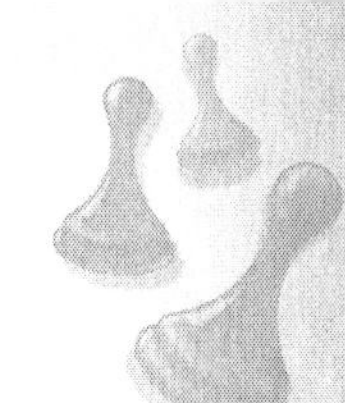

Pathway Puzzles

Cards *(cont.)*

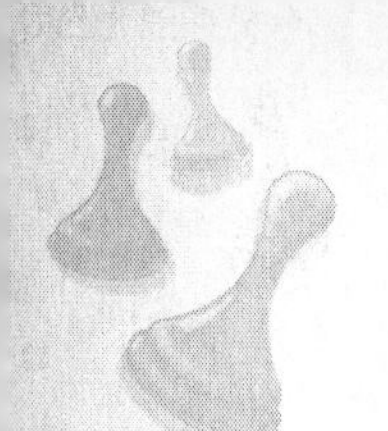

Pathway Puzzles

Cards *(cont.)*

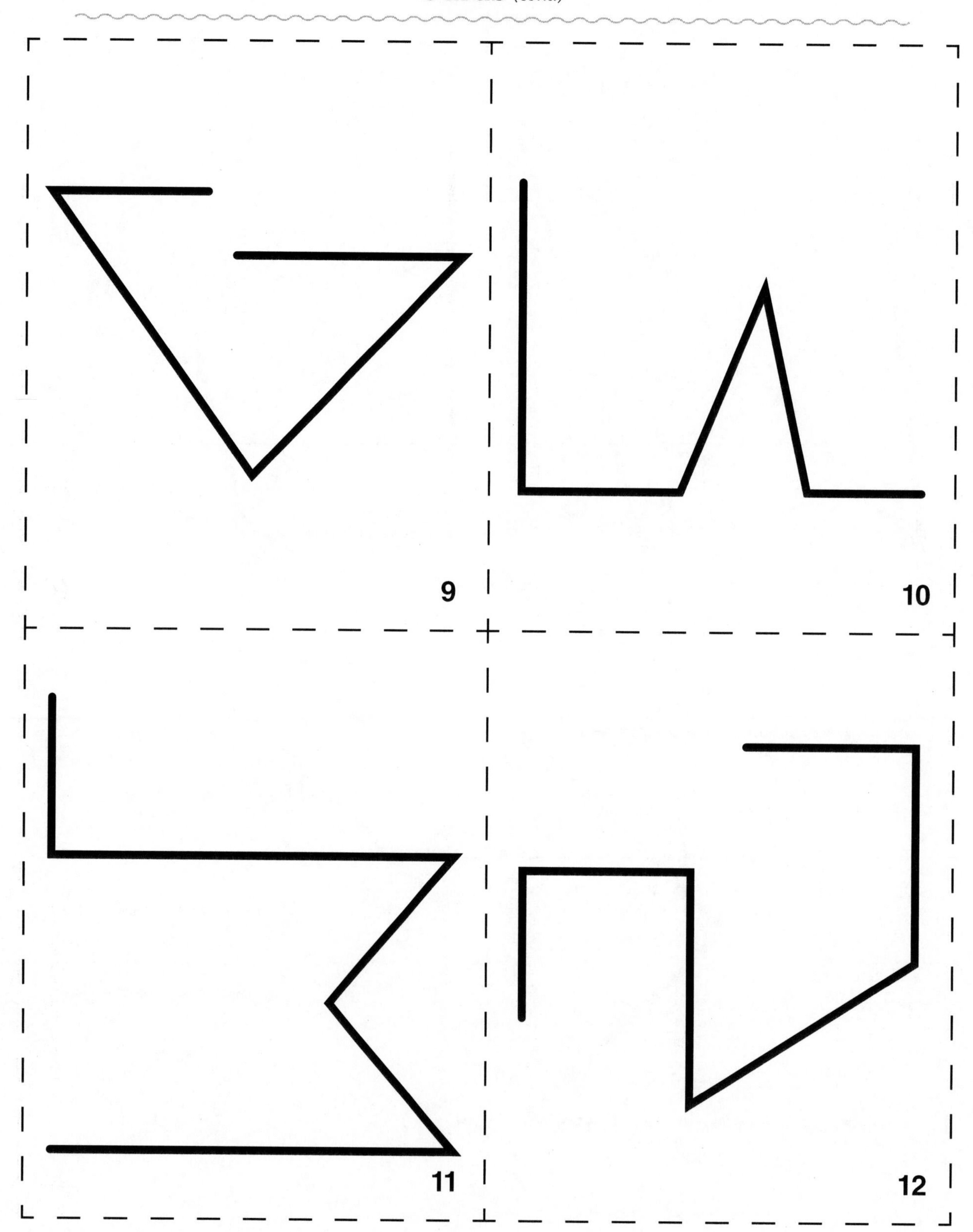

© Shell Education

Pathway Puzzles

Cards *(cont.)*

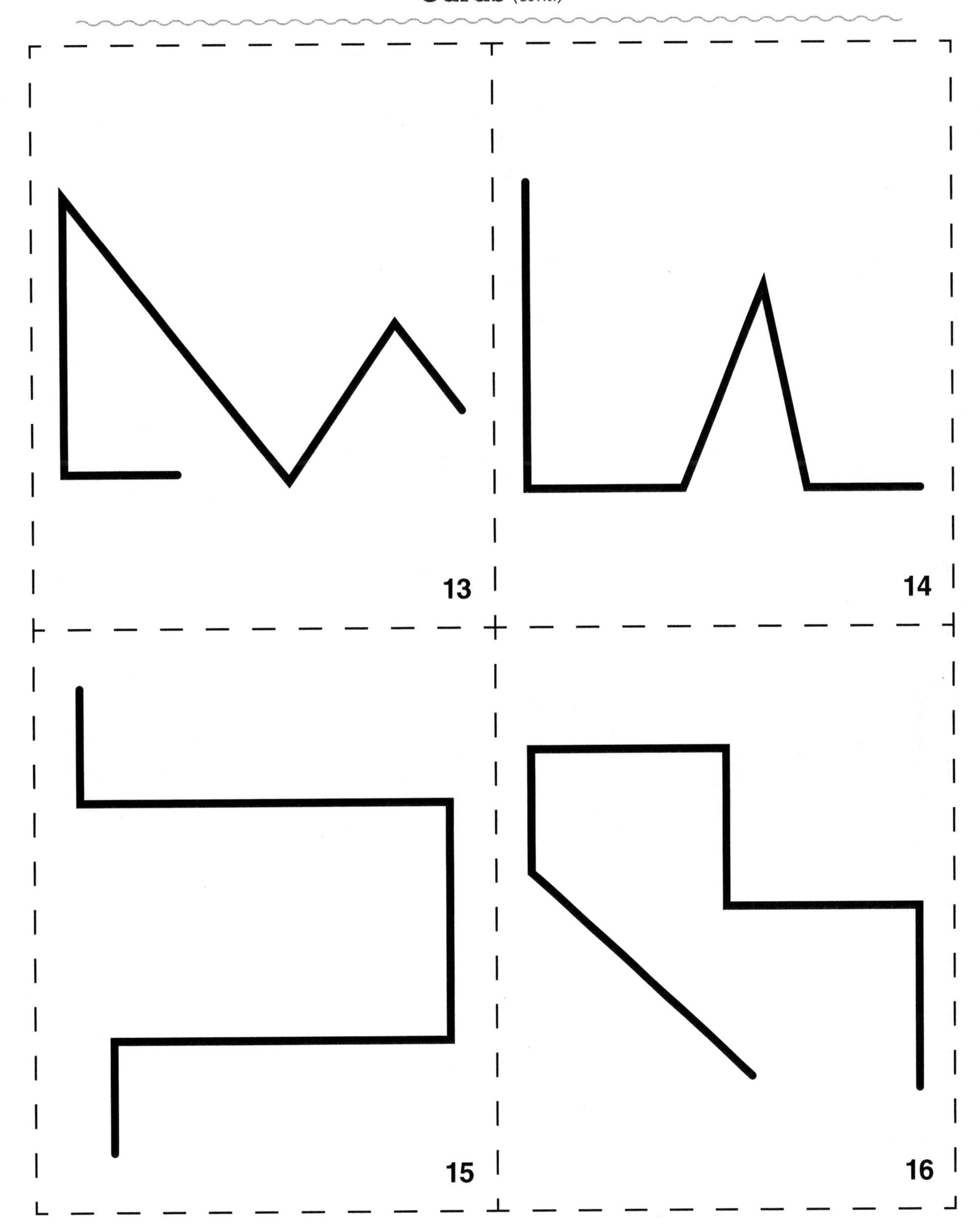

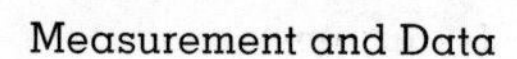

Name: ______________________________ Date: ________________

Pathway Puzzles

Recording Sheet

Directions: Record the length of each segment. Then convert the sum to millimeters.

Measurement of Segments in Centimeters: Sum of Segments in Millimeters:	Measurement of Segments in Centimeters: Sum of Segments in Millimeters:
Measurement of Segments in Centimeters: Sum of Segments in Millimeters:	Measurement of Segments in Centimeters: Sum of Segments in Millimeters:

 © Shell Education

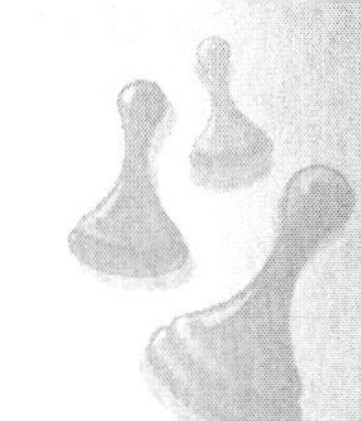

Pathway Puzzles
Game Board

Directions: Copy and cut out the game board. Tape it to the game board on page 120.

Pathway Puzzles

Game Board *(cont.)*

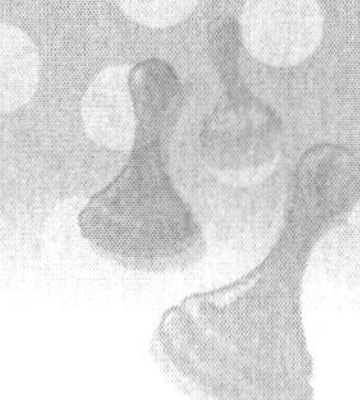

Measurement Match

Domain

Measurement and Data

Standard

Apply the area and perimeter formulas for rectangles in real world and mathematical problems.

Number of Players

Groups of 6 to 12

Materials

- *Measurement Match Cards* (pages 123–125)
- *Measurement Match Recording Sheet* (page 126)
- *Measurement Match Answer Key* (page 151)
- number cubes

GET PREPARED!

- Copy and cut out one set of the *Measurement Match Cards* for each group of players.
- Copy two *Measurement Match Recording Sheets* for each player.
- Provide the *Measurement Match Answer Key* for each player.
- Collect a number cube for each group of players.

Game Directions

1. Distribute materials to players.
2. Players take turns rolling a number cube. The player who rolls the lowest number is Player 1.
3. One player from each group acts as a dealer. He or she deals at least one card to each player. Depending on the group size, players may have more than one card.

Measurement Match *(cont.)*

4. Player 1 begins the game by reading the question on the right- hand side of the card aloud.

5. All players use their *Measurement Match Recording Sheet* to solve the problem and record the answer.

6. The player with the correct answer on the left side of his or her card reads it aloud. Answers are checked against the *Measurement Match Answer Key*. If the player answers the question correctly, he or she scores a point.

7. The player to the left of Player 1 goes next. He or she follows steps 4 to 6.

8. The game continues until all the questions have been answered. The player with the most points wins.

© Shell Education

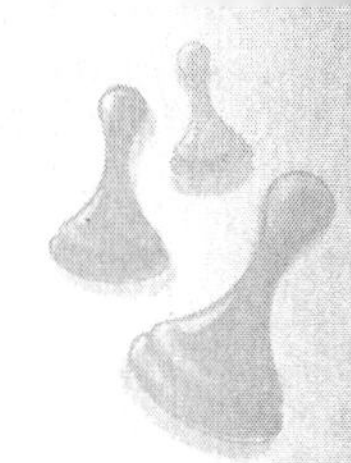

Measurement Match

Cards

Directions: Copy and cut out the cards for each group of players.

225 sq. ft.	**I am a rectangle with a length of 14 cm and a width of 6 cm.** **What is my perimeter?** **1**
40 cm	**I am a rectangle with an area of 56 square units. My length is 1 more unit than my width.** **What are my dimensions?** **2**
7 × 8 units	**I am a rectangle with a perimeter of 100 inches. My length is 28 inches.** **What is my width?** **3**
22 inches	**I am a square with a perimeter of 80 feet.** **What is my area?** **4**

Measurement Match

Cards *(cont.)*

400 sq. ft.	I am a rectangle with an area of 90 square feet. My length is 45 feet. What is my width? 5
2 feet	I am a rectangle with dimensions of 25 feet by 8 feet. What is my perimeter? 6
66 feet	I am a square with an area of 64 square feet. What is my perimeter? 7
32 feet	I am a square with a perimeter of 48 inches. What is my area? 8

© Shell Education

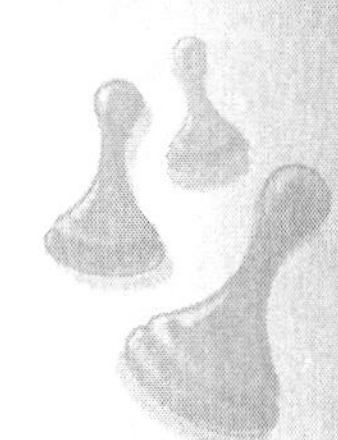

Measurement Match

Cards *(cont.)*

Answer	Question	Card
144 sq. in.	I am a rectangle with a perimeter of 50 feet. My length is 5 feet more than my width. What is my area?	9
150 sq. ft.	I am a square with a perimeter of 40 cm. What is my area?	10
100 sq. cm	I am a rectangle with an area of 42 square feet. My length is 1 foot more than my width. What is my perimeter?	11
26 feet	I am a square with perimeter of 60 feet. What is my area?	12

Name: ______________________ Date: ______________

Measurement Match

Recording Sheet

Directions: Record each problem in a box and solve it.

© Shell Education

Geo Identity

Domain

Geometry

Standard

Draw points, lines, line segments, rays, angles (right, acute, obtuse), and perpendicular and parallel lines. Identify these in two-dimensional figures.

Number of Players

2 Players

Materials

- *Geo Identity Cards* (pages 129–132)
- number cubes

GET PREPARED!

- Copy and cut out the *Geo Identity Cards* for each pair of players.
- Collect a number cube for each pair of players.

Game Directions

1. Distribute materials to players.
2. Players take turns rolling a number cube. The player who rolls the lower number is Player 1.
3. Player 1 deals five cards to each player.

Geo Identity *(cont.)*

4. The remaining cards are placed facedown between the two players. The top card is turned over to begin a discard pile.

5. Player 2 draws the top card from the deck or the discard pile. The player checks the cards in his or her hand to find a pair. A pair is a figure card and a matching description card. (**Note:** There are three pairs for every description.) For example:

6. If the player has a pair, he or she places the two cards in his or her winning pile and discards one card. If the player does not have a pair, he or she just discards one card.

7. When the deck is depleted, the discard pile is shuffled and placed facedown. The game continues until all the cards have been played. The player with more pairs wins the game.

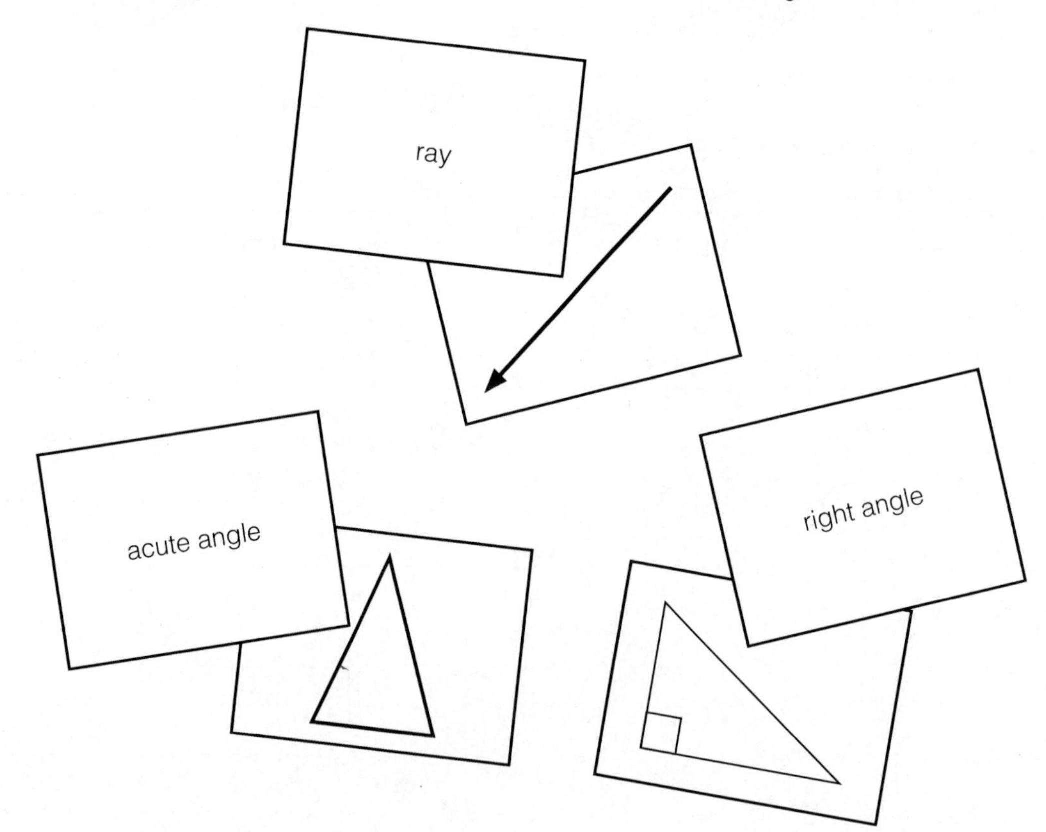

© Shell Education

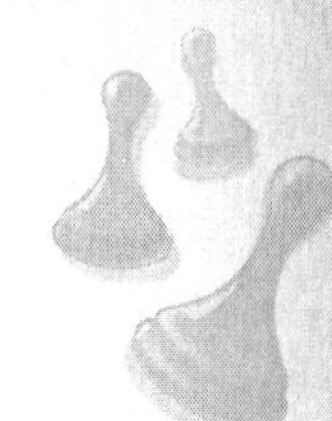

Geo Identity

Cards

Directions: Copy and cut out the cards for each group of players.

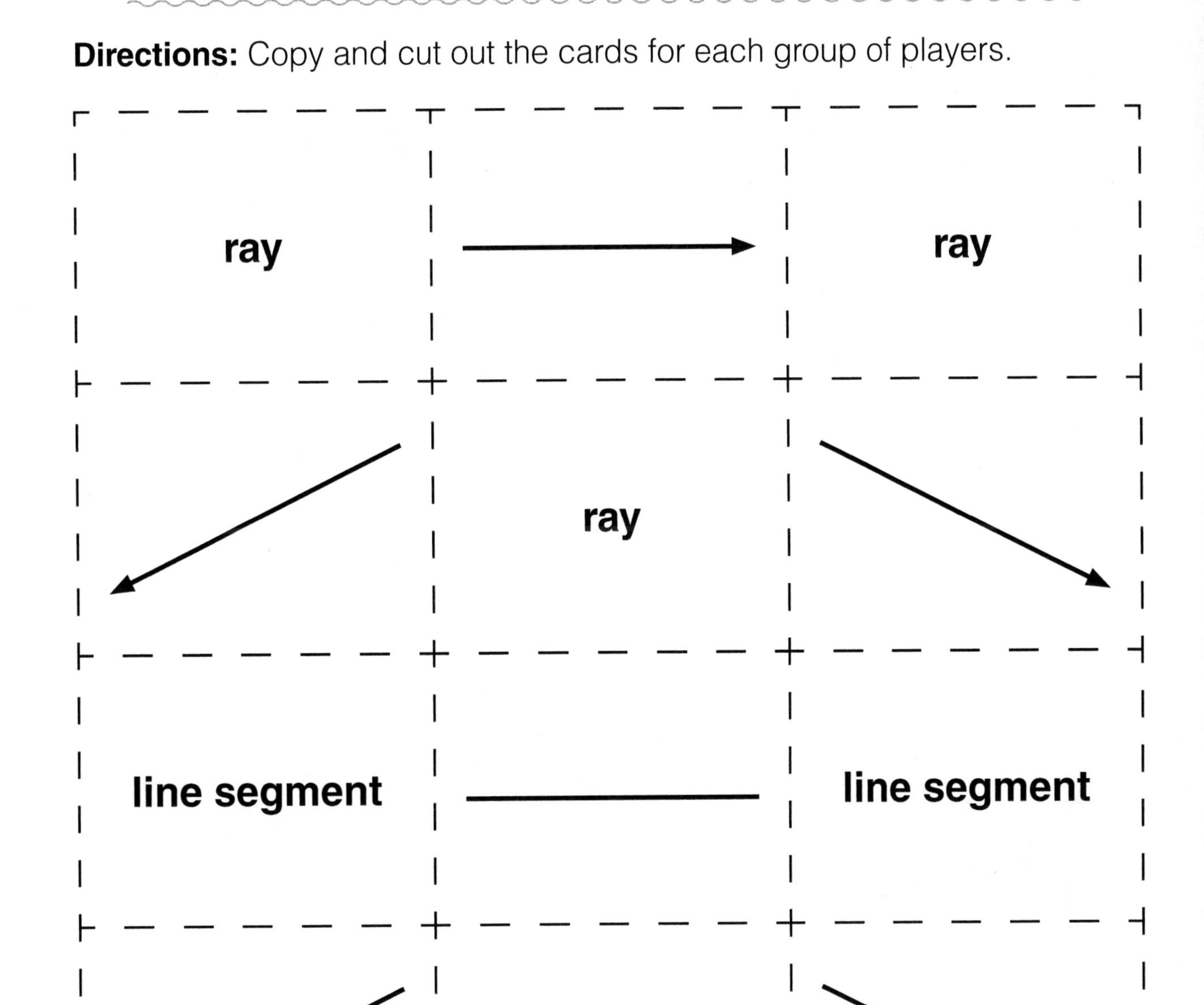

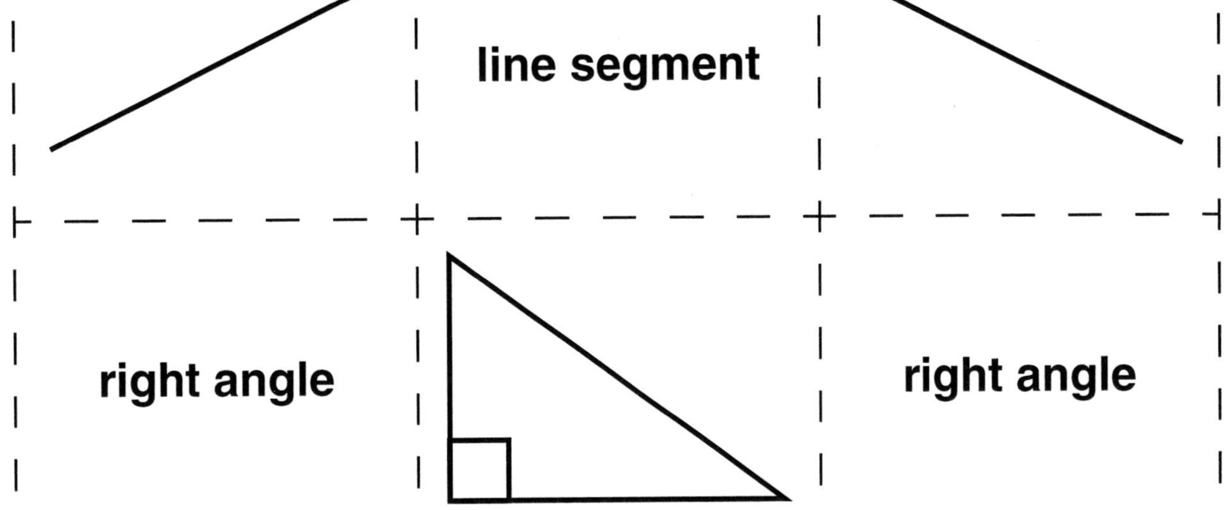

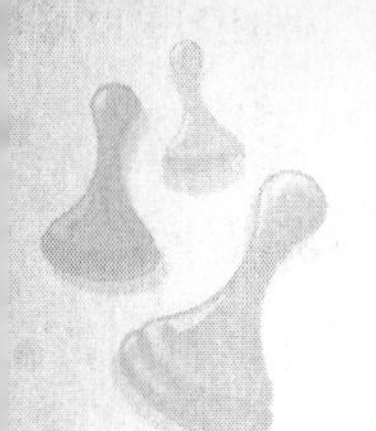

Geo Identity

Cards *(cont.)*

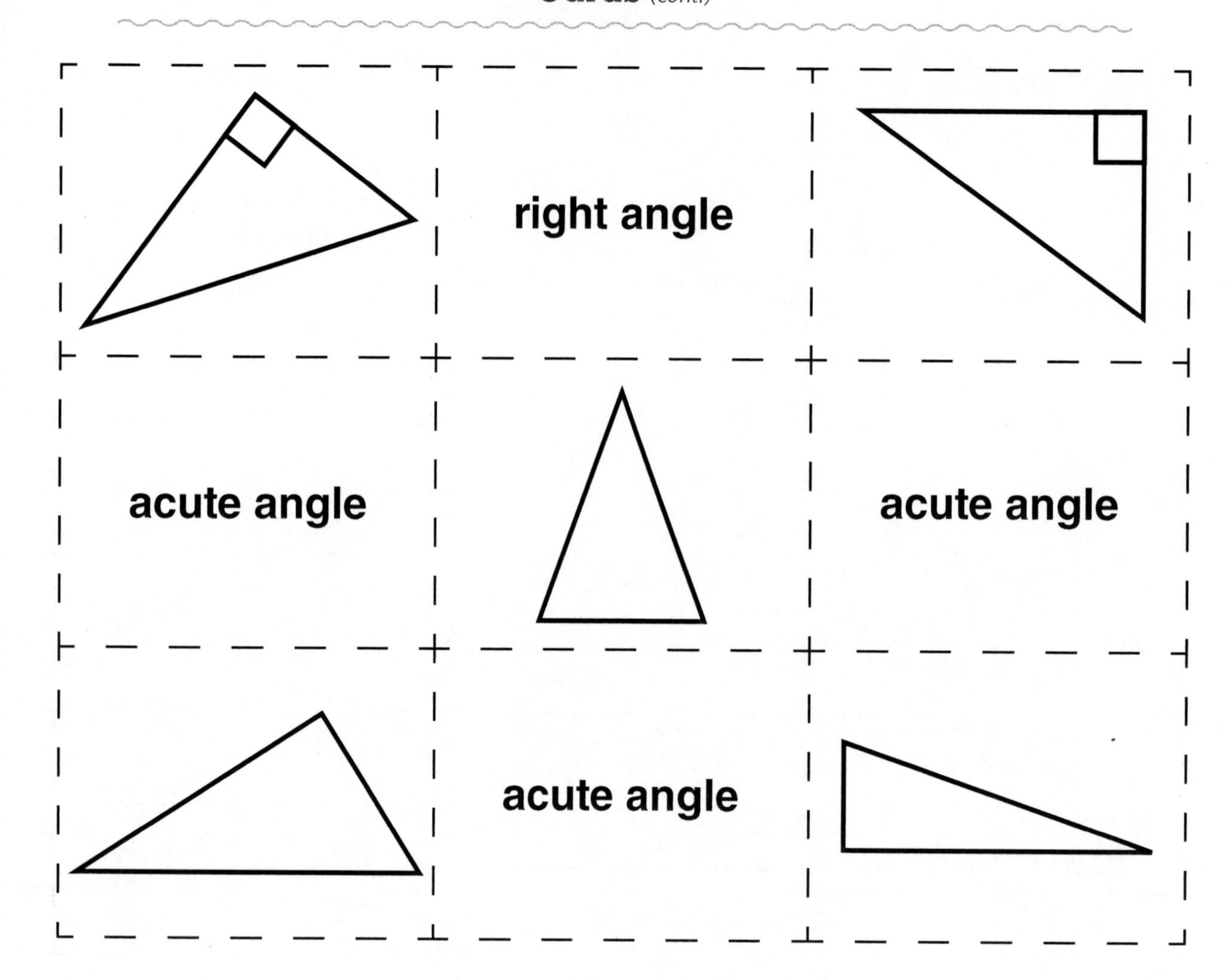

© Shell Education

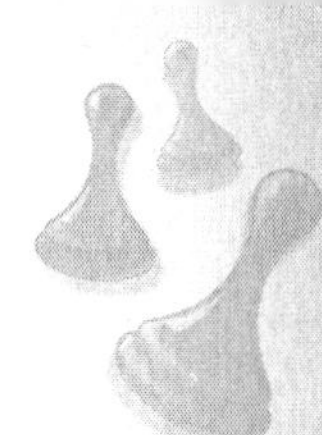

Geo Identity

Cards *(cont.)*

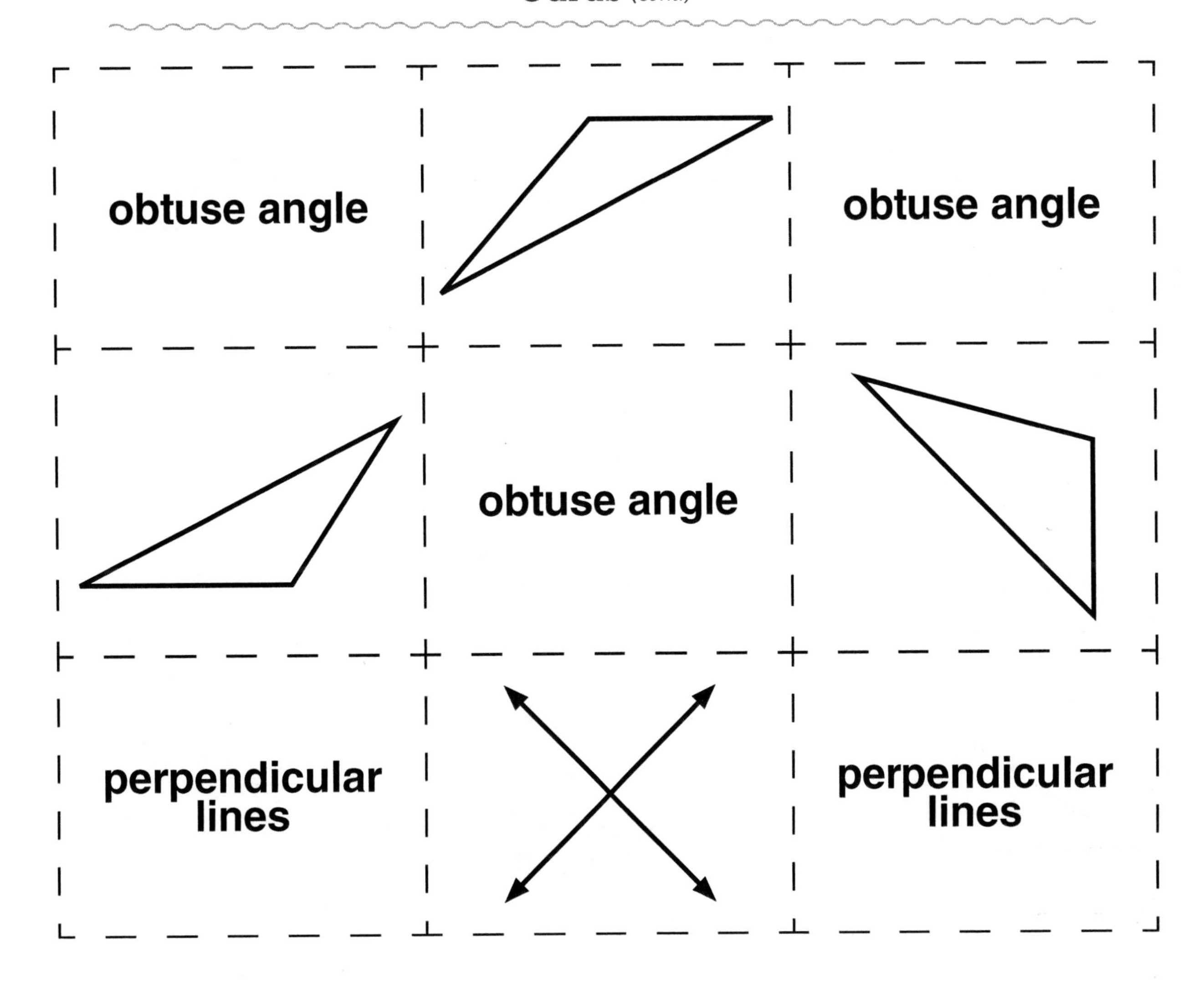

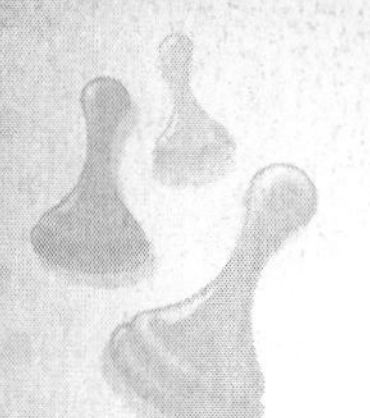

Geo Identity

Cards *(cont.)*

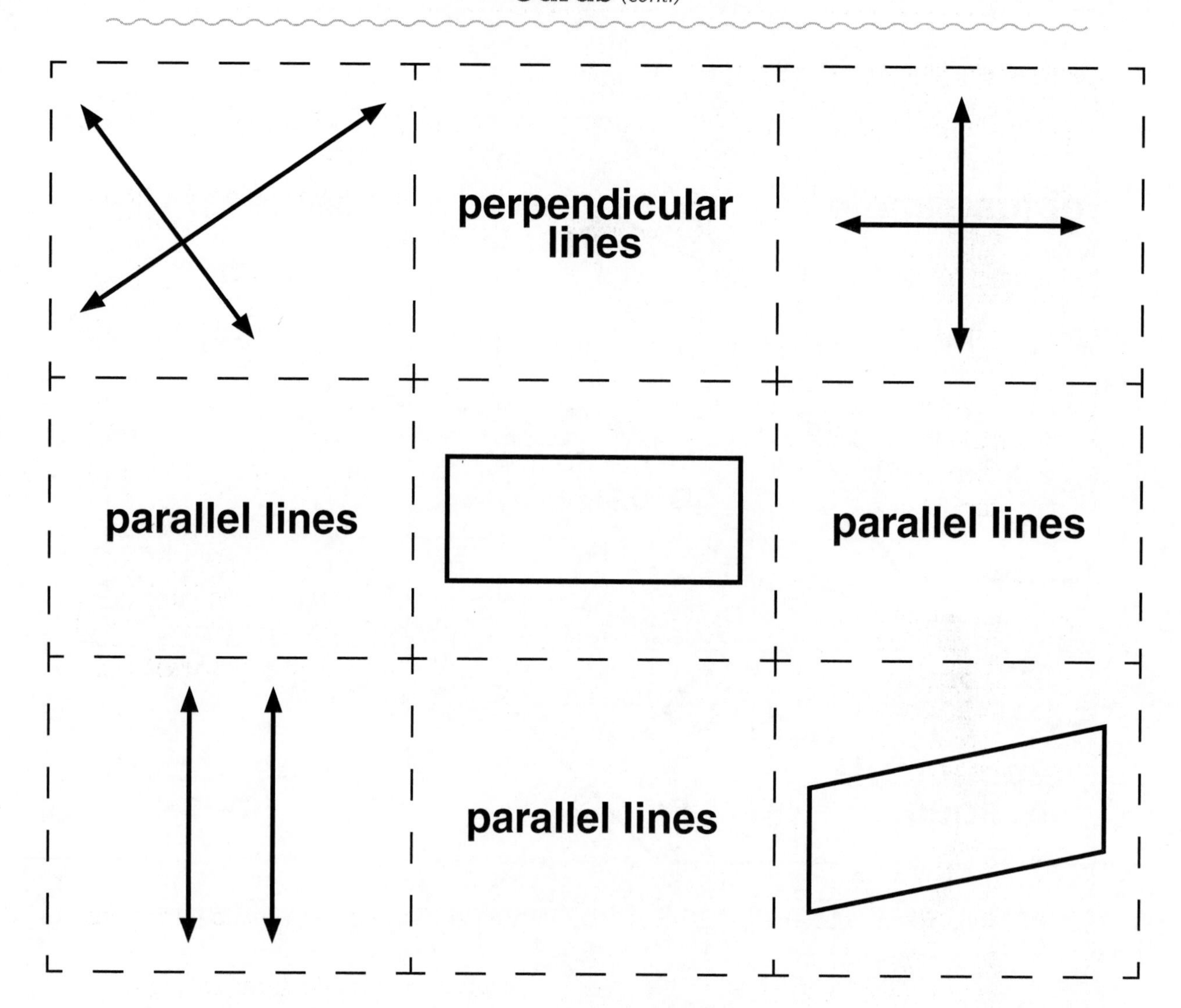

© Shell Education

Symmetric Shapes

Domain

Geometry

Standard

Recognize a line of symmetry for a two-dimensional figure as a line across the figure such that the figure can be folded along the line into matching parts. Identify line-symmetric figures and draw lines of symmetry.

Number of Players

2 Players

Materials

- *Symmetric Shapes Cards* (pages 135–136)
- *Symmetric Shapes Number Cards* (pages 137–138)
- *Symmetric Shapes Recording Sheet* (page 139)
- *Symmetric Shapes Answer Key* (page 152)
- rulers
- number cubes

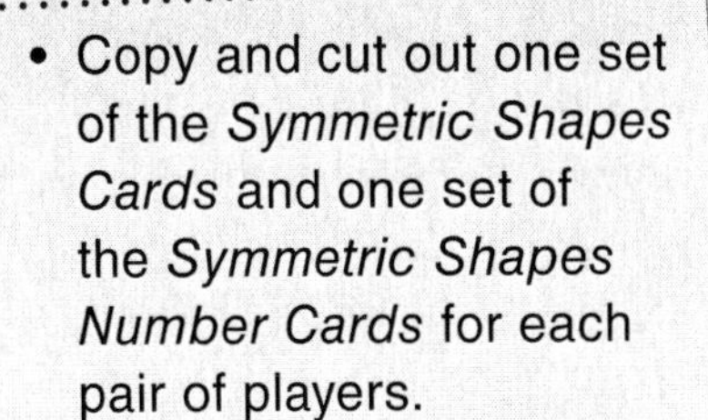

GET PREPARED!

- Copy and cut out one set of the *Symmetric Shapes Cards* and one set of the *Symmetric Shapes Number Cards* for each pair of players.
- Make two copies of the *Symmetric Shapes Recording Sheet* for each player.
- Collect a ruler and a number cube for each pair of players.

Game Directions

1. Distribute materials to players.
2. Players take turns rolling a number cube. The player who rolls the lower number is Player 1.
3. One player shuffles the shape cards, places them facedown, and deals 12 *Symmetric Shapes Cards* to each player.

Symmetric Shapes *(cont.)*

4. Players draw all possible lines of symmetry for each of the shapes on their *Symmetric Shapes Recording Sheet*. They jot down the number of lines on the card.

5. Player 1 draws one of the *Symmetric Shapes Number Cards* from the deck. The numbers indicate the number of lines of symmetry in a shape. If either player can match a geometric figure with the number of lines of symmetry, the player places the matching shape cards on top of the number card. If a match can be made by either player, the next player draws a *Symmetric Shapes Number Card.*

6. Player 2 draws a number card. If either player makes a match, he or she places the shape card on the desk.

7. If the number card deck is depleted, it is reshuffled. The game continues until the first player plays all of his or her shape cards. The player with more matches is the winner.

 © Shell Education

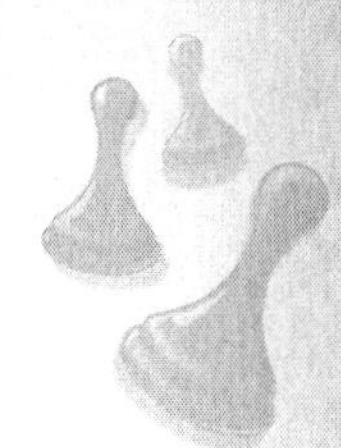

SYMMETRIC SHAPES

Cards

Directions: Copy and cut out one set of cards for each pair of players.

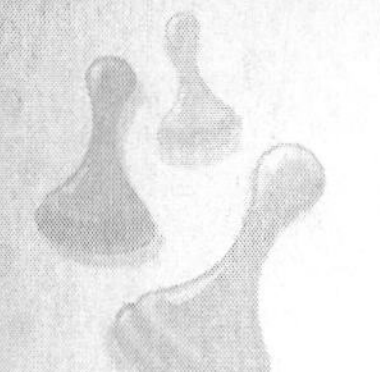

SYMMETRIC SHAPES

Cards *(cont.)*

 © Shell Education

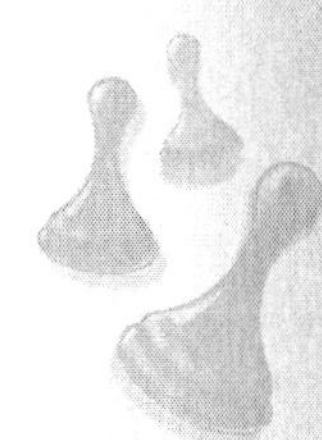

SYMMETRIC SHAPES

Number Cards

Directions: Copy and cut out the cards for each group of players.

1	1	1	1
1	1	1	1
1	1	2	2

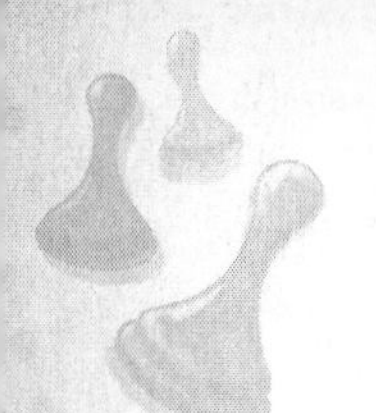

SYMMETRIC SHAPES

Number Cards *(cont.)*

2	2	2	2
2	3	4	4
2	5	6	8

© Shell Education

Name: ______________________________ Date: ________________

SYMMETRIC SHAPES

Recording Sheet

Directions: Use the recording sheets to draw the shapes and lines of symmetry.

Classify Bingo

Domain

Geometry

Standard

Classify two-dimensional figures based on the presence or absence of parallel or perpendicular lines, or the presence or absence of angles of a specified size. Recognize right triangles as a category, and identify right triangles.

Number of Players

Groups of 2 to 4

Materials

- *Classify Bingo Boards* (pages 141–144)
- *Classify Bingo Cards* (pages 145–147)
- bingo markers (e.g., bingo chips, colored counters, or coins)
- number cubes

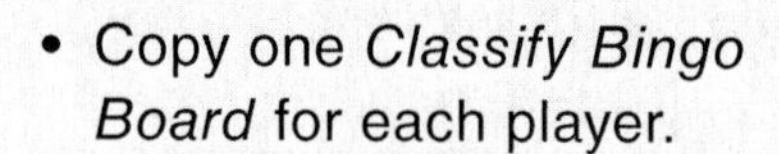

- Copy one *Classify Bingo Board* for each player.
- Copy and cut out one set of the *Classify Bingo Cards* for each group of players.
- Collect bingo markers and a number cube for each pair of players.

Game Directions

1. Distribute materials to players.
2. Players take turns rolling a number cube. The player who rolls the higher number is Player 1. If more than two players roll the number cube, player number is determined from highest to lowest roll.
3. Player 1 begins the game by turning over the top card. All players mark the corresponding space on their bingo boards.
4. The rest of the players repeat Step 3. All players mark the corresponding space on their bingo boards.
5. The first player in the group to mark three spaces in a row—horizontally, vertically, or diagonally—calls "Bingo!" If the responses are correct, he or she wins the game.

© Shell Education

Name: ______________________ Date: ______________

Classify Bingo

Board 1

Directions: Mark the space on the board to connect three in a row.

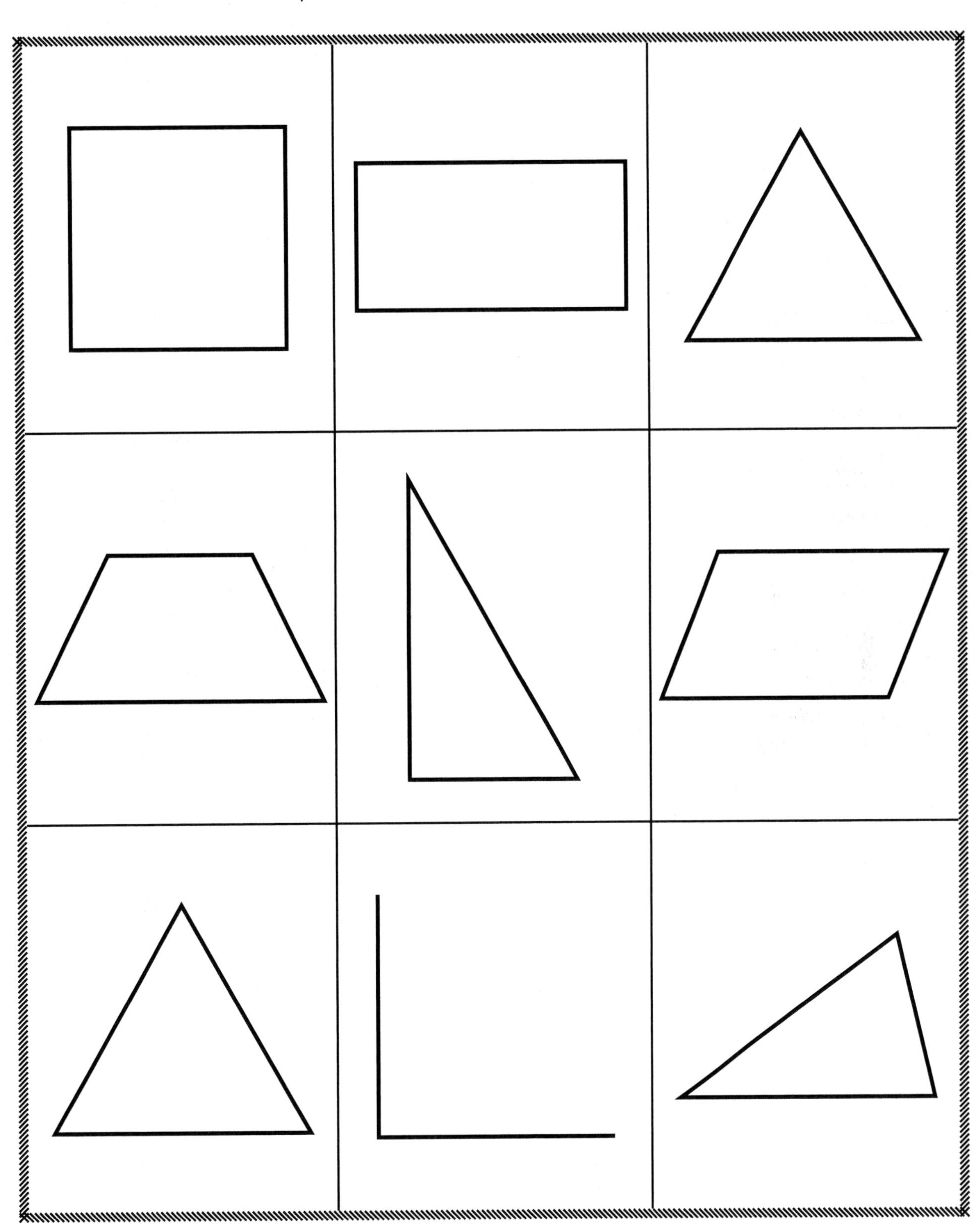

© Shell Education

Name: ______________________ Date: ______________

Classify Bingo

Board 2

Directions: Mark the space on the board to connect three in a row.

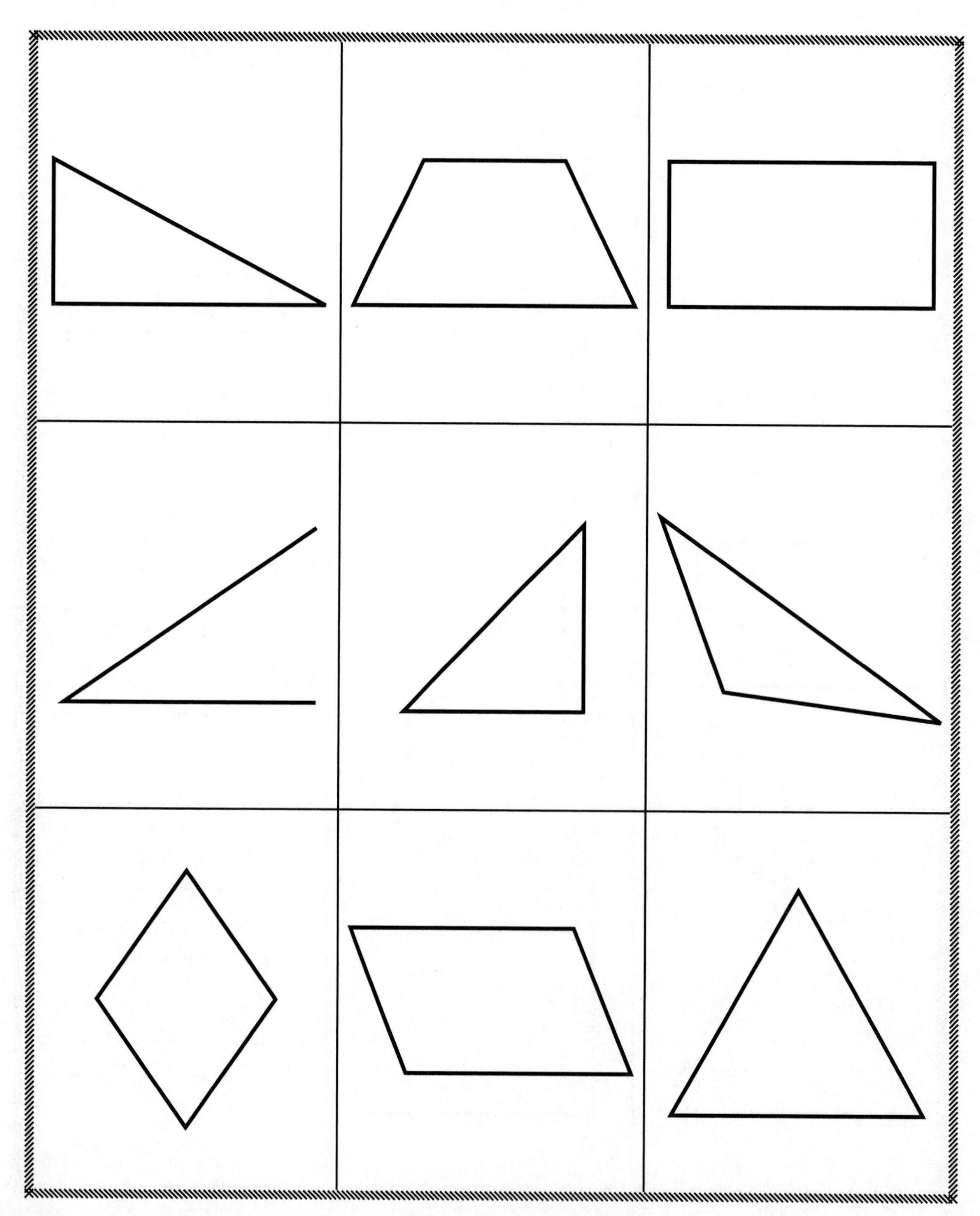

© Shell Education

Name:__________________________________ Date:____________________

Classify Bingo

Board 3

Directions: Mark the space on the board to connect three in a row.

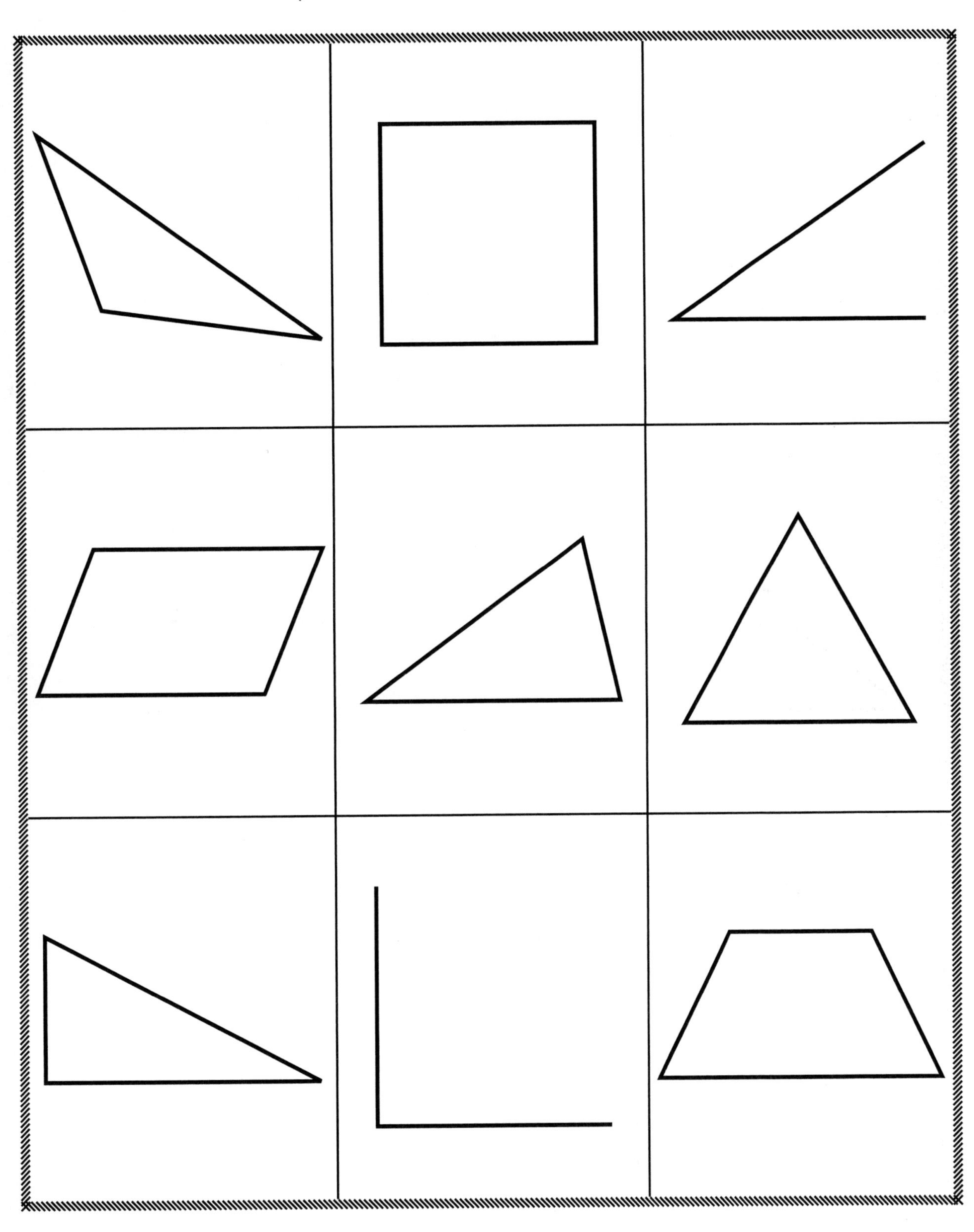

© Shell Education

Name:______________________________ Date:________________

Classify Bingo

Board 4

Directions: Mark the space on the board to connect three in a row.

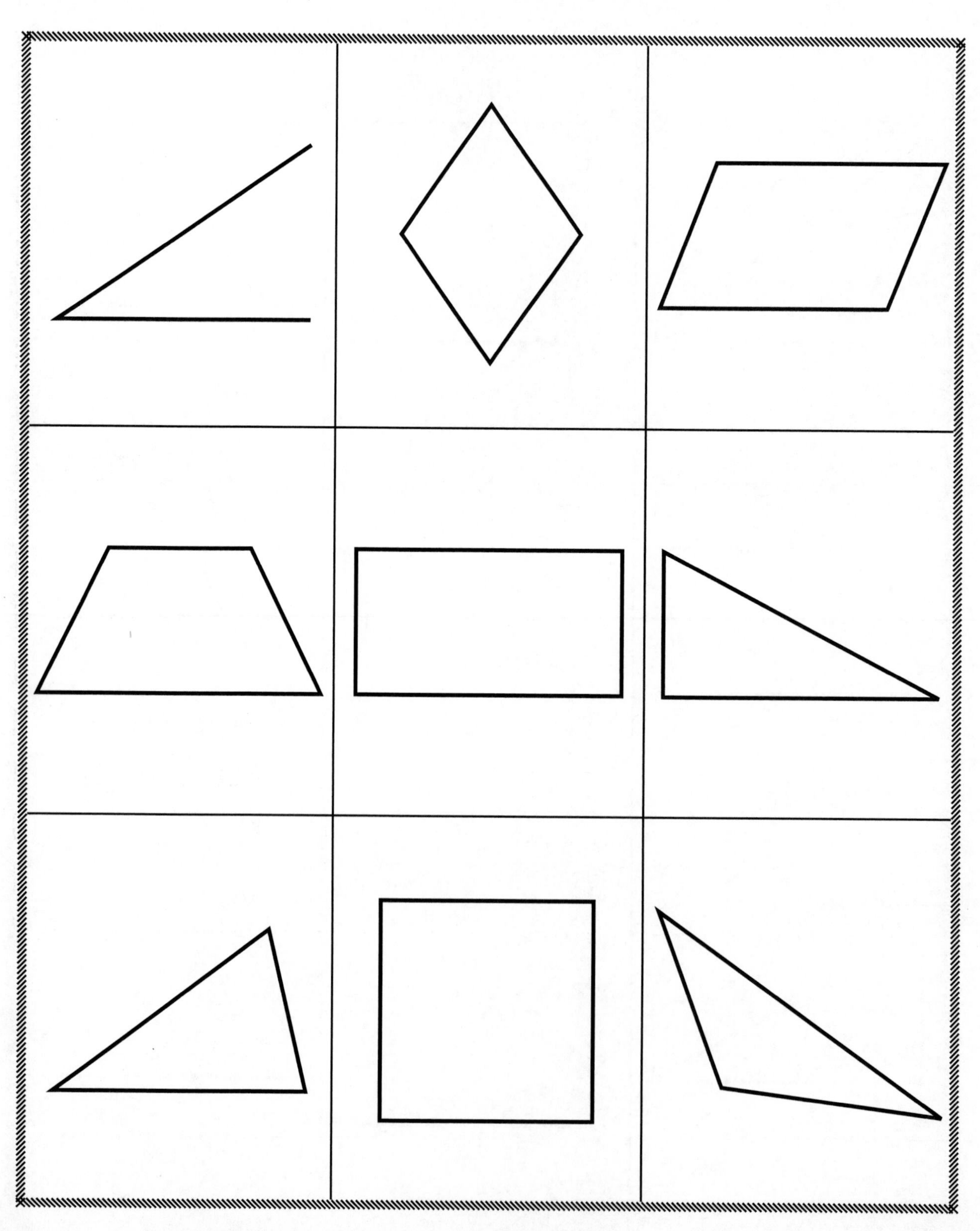

 © Shell Education

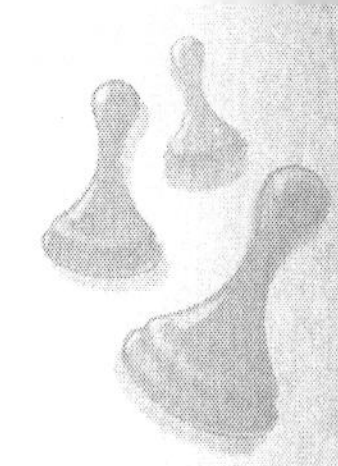

Classify Bingo
Cards

Directions: Copy and cut out the cards for each group of players.

Right Triangle	**Right Triangle**	**Obtuse Triangle**
Obtuse Triangle	**Acute Triangle**	**Acute Triangle**
Right Angle	**Right Angle**	**Acute Angle**

© Shell Education

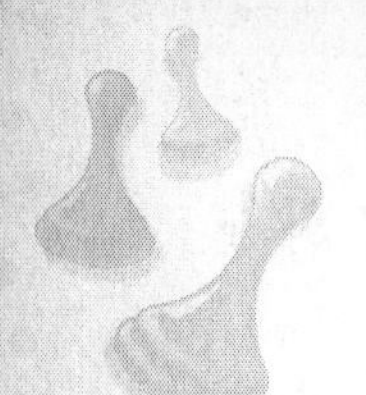

Classify Bingo

Cards *(cont.)*

Acute Angle	**Obtuse Angle**	**Obtuse Angle**
Parallel Lines	**Parallel Lines**	**Perpendicular Lines**
Perpendicular Lines	**Sum of Angles is 180 Degrees**	**Sum of Angles is 180 Degrees**

© Shell Education

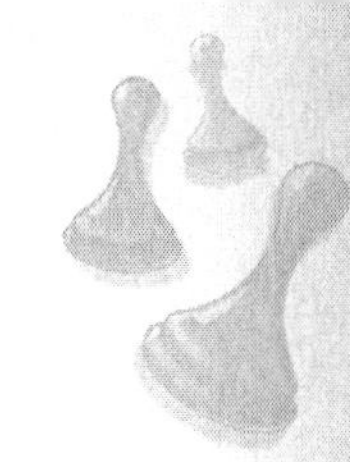

Classify Bingo

Cards *(cont.)*

Sum of Angles is 360 Degrees	Sum of Angles is 360 Degrees	Four Right Angles
Four Right Angles	60 Degree Angle	60 Degree Angle
Parallel Sides	Parallel Sides	90 Degree Angle
90 Degree Angle	Right Angle	Right Angle

© Shell Education

References Cited

Burns, Marilyn. 2009. "Win-Win Math Games." *Instructor.* Reprinted March/April, http://www.mathsolutions.com/documents/winwin_mathgames.pdf.

Hull, Ted H., Ruth Harbin Miles, and Don S. Balka. 2013. *Math Games: Getting to the Core of Conceptual Understanding.* Huntington Beach, CA: Shell Education.

National Council of Teachers of Mathematics. 2000. *Principles and Standards for School Mathematics.* Reston, VA: NCTM.

National Governors Association Center for Best Practices, and Council of Chief State School Officers. 2010. "Common Core State Standards." Washington, DC: National Governors Association Center for Best Practices, Council of Chief State School Officers. Accessed September 23, 2013, http://corestandards.org/math.

National Research Council. 2001. "Adding It Up: Helping Children Learn Mathematics." Washington, DC: National Academy Press.

National Research Council. 2004. "Engaging Schools: Fostering High School Students' Motivation to Learn." Washington, DC: National Academy Press.

 © Shell Education

Contents of the Digital Resource CD

Student Resources		
Page(s)	Title	Filename
18–21	Factor Frenzy Cards	frenzycards.pdf
22	Factor Frenzy Game Sheet	frenzysheet.pdf
23	Factor Frenzy Grid	frenzygrid.pdf
26–28	Factor Pairs Number Cards	paircards.pdf
29	Factor Pairs Recording Sheet Part 1	pairsheet1.pdf
30	Factor Pairs Recording Sheet Part 2	pairsheet2.pdf
33	Pattern Poser Rule Cards	patterncards.pdf
34	Pattern Poser Recording Sheet	patternsheet.pdf
37–38	Multiplication Sprint Game Board	sprintboard.pdf
39	Multiplication Sprint Game Markers	sprintmarkers.pdf
40–41	Multiplication Sprint Recording Sheet	sprintsheet.pdf
42	Multiplication Sprint Digit Deck	sprintdeck.pdf
45	Spin and Round Recording Sheet	spinsheet.pdf
46–47	Spin and Round Digit Deck	spindeck.pdf
48	Spin and Round Spinner	spinspinner.pdf
51–52	Standing Order Number Cards Set 1	standingset1.pdf
53–54	Standing Order Number Cards Set 2	standingset2.pdf
55–56	Standing Order Number Cards Set 3	standingset3.pdf
57–58	Standing Order Number Cards Set 4	standingset4.pdf
59–60	Standing Order Number Cards Set 5	standingset5.pdf
61–62	Standing Order Number Cards Set 6	standingset6.pdf
65–66	Remainder Race Game Board	raceboard.pdf
67	Remainder Race Game Markers	racemarker.pdf
70–75	What's My Name? Number Cards	namecard.pdf
78–80	Fishing for Fractions Cards	fishingcard.pdf
83	Dueling Decimals Spinner	duelingspinner.pdf
84	Dueling Decimals Recording Sheet	duelingsheet.pdf
87–89	Fraction Flip Cards	fractioncard.pdf

Contents of the Digital Resource CD *(cont.)*

Student Resources		
Page(s)	**Title**	**Filename**
90	Fraction Flip Strips	fractionstrip.pdf
93–96	Fraction Decimal Match Cards	fractionmatch.pdf
99	Equivalent Fractions Bingo Board	equivalentbingo.pdf
100–101	Equivalent Fractions Cards	equivalentcard.pdf
104–105	Fraction Action Race Game Board	actionboard.pdf
106	Fraction Action Race Fraction Strips	actionstrip.pdf
109	Improper Fraction Bingo Board	improperboard.pdf
110	Improper Fraction Bingo Spinner	improperspinner.pdf
111	Improper Fraction Recording Sheet	impropersheet.pdf
114–117	Pathway Puzzles Cards	puzzlecards.pdf
118	Pathway Puzzles Recording Sheet	puzzlesheet.pdf
119 –120	Pathway Puzzles Game Board	puzzleboard.pdf
123–125	Measurement Match Cards	measurecard.pdf
126	Measurement Match Recording Sheet	measuresheet.pdf
129–132	Geo Identity Cards	geocards.pdf
135–136	Symmetric Shapes Cards	shapecards.pdf
137–138	Symmetric Shapes Number Cards	numbercards.pdf
139	Symmetric Shapes Recording Sheet	shapessheet.pdf
141–144	Classify Bingo Boards	classifyboard.pdf
145–147	Classify Bingo Cards	classifycard.pdf

Additional Resources	
Title	**Filename**
CCSS, WIDA, and TESOL	standards.pdf

© Shell Education

Answer Key

Pathway Puzzles (pages 114–117)

1. 95 mm
2. 95 mm
3. 157.5 mm
4. 90 mm
5. 240 mm
6. 240 mm
7. 240 mm
8. 155 mm
9. 180 mm
10. 177.5 mm
11. 242.5 mm
12. 210 mm
13. 190 mm
14. 177.5 mm
15. 207.5 mm
16. 205 mm

Measurement Match (pages 123–125)

1. 40 cm
2. 7×8 units
3. 22 inches
4. 400 sq. ft.
5. 2 feet
6. 66 feet
7. 32 feet
8. 144 sq. in.
9. 150 sq. cm.
10. 100 sq. cm.
11. 26 feet
12. 225 sq. ft.

Answer Key *(cont.)*

Symmetric Shapes (pages 135–136)

1. 4 lines of symmetry

2. 1 line of symmetry

3. 2 lines of symmetry

4. 3 lines of symmetry

5. 4 lines of symmetry

6. 1 line of symmetry

7. 1 line of symmetry

8. 2 lines of symmetry

9. 1 line of symmetry

10. 5 lines of symmetry

11. 6 lines of symmetry

12. 2 lines of symmetry

13. 1 line of symmetry

14. 1 line of symmetry

15. 1 line of symmetry

16. 2 lines of symmetry

17. 8 lines of symmetry

18. 2 lines of symmetry

19. 2 lines of symmetry

20. 2 lines of symmetry

21. 1 line of symmetry

22. 1 line of symmetry

23. 2 lines of symmetry

24. 1 line of symmetry

© Shell Education